LONDON MATHEMATICAL SOCIETY LECTURE NOTE SERIES

Managing Editor: Professor I.M.Jame
Mathematical Institute, 24-29 St Gil

1. General cohomology theory and K-
4. Algebraic topology: a student's
5. Commutative algebra, J.T.KNIGHT
8. Integration and harmonic analysis on compact groups, R.E.EDWARDS
9. Elliptic functions and elliptic curves, P.DU VAL
10. Numerical ranges II, F.F.BONSALL and J.DUNCAN
11. New developments in topology, G.SEGAL (ed.)
12. Symposium on complex analysis, Canterbury, 1973, J.CLUNIE and W.K.HAYMAN (eds.)
13. Combinatorics: Proceedings of the British combinatorial conference 1973, T.P.McDONOUGH and V.C.MAVRON (eds.)
14. Analytic theory of abelian varieties, H.P.F.SWINNERTON-DYER
15. An introduction to topological groups, P.J.HIGGINS
16. Topics in finite groups, T.M.GAGEN
17. Differentiable germs and catastrophes, Th.BROCKER and L.LANDER
18. A geometric approach to homology theory, S.BUONCRISTIANO, C.P.ROURKE and B.J.SANDERSON
20. Sheaf theory, B.R.TENNISON
21. Automatic continuity of linear operators, A.M.SINCLAIR
23. Parallelisms of complete designs, P.J.CAMERON
24. The topology of Stiefel manifolds, I.M.JAMES
25. Lie groups and compact groups, J.F.PRICE
26. Transformation groups: Proceedings of the conference in the University of Newcastle upon Tyne, August 1976, C.KOSNIOWSKI
27. Skew field constructions, P.M.COHN
28. Brownian motion, Hardy spaces and bounded mean oscillation, K.E.PETERSEN
29. Pontryagin duality and the structure of locally compact abelian groups, S.A.MORRIS
30. Interaction models, N.L.BIGGS
31. Continuous crossed products and type III von Neumann algebras, A.VAN DAELE
32. Uniform algebras and Jensen measures, T.W.GAMELIN
33. Permutation groups and combinatorial structures, N.L.BIGGS and A.T.WHITE
34. Representation theory of Lie groups, M.F.ATIYAH et al.
35. Trace ideals and their applications, B.SIMON
36. Homological group theory, C.T.C.WALL (ed.)
37. Partially ordered rings and semi-algebraic geometry, G.W.BRUMFIEL
38. Surveys in combinatorics, B.BOLLOBAS (ed.)
39. Affine sets and affine groups, D.G.NORTHCOTT
40. Introduction to H_p spaces, P.J.KOOSIS
41. Theory and application of Hopf bifurcation, B.D.HASSARD, N.D.KAZARINOFF and Y-H.WAN
42. Topics in the theory of group presentations, D.L.JOHNSON
43. Graphs, codes and designs, P.J.CAMERON and J.H.VAN LINT
44. Z/2-homotopy theory, M.C.CRABB
45. Recursion theory: its generalisations and applications, F.R.DRAKE and S.S.WAINER (eds.)
46. p-adic analysis: a short course on recent work, N.KOBLITZ

London Mathematical Society Lecture Note Series. 46

p-adic Analysis:
a Short Course on Recent Work

Neal Koblitz
Assistant Professor of Mathematics
University of Washington

CAMBRIDGE UNIVERSITY PRESS

CAMBRIDGE

LONDON NEW YORK ROCHELLE

MELBOURNE SYDNEY

Published by the Press Syndicate of the University of Cambridge
The Pitt Building, Trumpington Street, Cambridge CB2 1RP
32 East 57th Street, New York, NY 10022, USA
296 Beaconsfield Parade, Middle Park, Melbourne 3206, Australia

First published 1980

British Library cataloguing in publication data

Koblitz, Neal
P-adic analysis. - (London Mathematical Society
Lecture Note Series; 46 ISSN 0076-0552).
1. p-adic numbers
I. Title II. Series
512'.74 QA241 80-40806

ISBN 0 521 28060 5

Transferred to digital printing 2004

CONTENTS

PREFACE

This book is a revised and expanded version of a series of talks given in Hanoi at the Viện Toán học (Mathematical Institute) in July, 1978. The purpose of the book is the same as the purpose of the talks: to make certain recent applications of p-adic analysis to number theory accessible to graduate students and researchers in related fields. The emphasis is on new results and conjectures, or new interpretations of earlier results, which have come to light in the past couple of years and which indicate intriguing and as yet imperfectly understood new connections between algebraic number theory, algebraic geometry, and p-adic analysis.

I occasionally state without proof or assume some familiarity with facts or techniques of other fields: algebraic geometry (Chapter III), algebraic number theory (Chapter IV), analysis (the Appendix). But I include down-to-earth examples and words of motivation whenever possible, so that even a reader with little background in these areas should be able to see what's going on.

Chapter I contains the basic information about p-adic numbers and p-adic analysis needed for what follows. Chapter II describes the construction and properties of p-adic Dirichlet L-functions, including Leopoldt's formula for the value at 1, using the approach of p-adic integration. The p-adic gamma function and log gamma function are introduced, their properties are developed and compared with the identities satisfied by the classical gamma function, and two formulas relating them to the p-adic L-functions $L_p(s,\chi)$ are proved. The first formula--expressing $L'_p(0,\chi)$ in terms of special values of log gamma--will be used later (Chapter IV) in the discussion of Gross' p-adic regulator. The other formula--a p-adic Stirling series for log gamma near infinity--will be a key motivating example for the p-adic Stieltjes transform, discussed in the Appendix.

Chapter III is devoted primarily to proving a p-adic formula for Gauss sums, which expresses them essentially as values of the p-adic gamma function. The approach emphasizes the analogy with the complex-analytic periods of differentials on certain special curves, and uses some algebraic geometry. The reader who is interested in a treatment that is more "elementary" and self-contained (but more computational rather than geometric) is referred to [62].

Chapter IV discusses two different types of p-adic regulators. One, due to Leopoldt, is connected with the behavior of $L_p(s,\chi)$ at $s=1$; the other, due to Gross, is connected with the behavior at $s=0$. Conjectures describing these connections between regulators and L-functions are explained and compared to the classical case. The conjectures are proved in the case of a one-dimensional character χ with base field Q (the "abelian over Q" case). The proof of Gross' conjecture in this case combines the formula for $L_p'(0,\chi)$ in Chapter II and the p-adic formula for Gauss sums in Chapter III, together with a p-adic version of the linear independence over $\overline{Q}$ of logarithms of algebraic numbers (Baker's theorem). This proof provides the culmination of the main part of the book.

The Appendix concerns some general constructions in p-adic analysis: the Stieltjes transform and the Shnirelman integral. I first use the Stieltjes transform to highlight the analogy between the p-adic and classical log gamma functions. I then give a complete account of M. M. Vishik's p-adic spectral theorem. This material has been relegated to the Appendix because it has not yet led to new number theoretic or algebra-geometric facts, perhaps because Vishik's theory is not very well known.

I would like to thank N. M. Katz, whose Spring 1978 lectures at

Princeton provided the explanations of the algebraic geometry and p-adic cohomology given in Chapter III; R. Greenberg, whose seminar talks at the University of Washington in October 1979 and whose comments on the manuscript were of great help in writing Chapter IV; B. H. Gross, whose preprint [35] and correspondence were the basis for the second half of Chapter IV; and M. M. Vishik, whose preprint [95] is given in modified form in §§3-4 of the Appendix.

I am also grateful to Ju. I. Manin and A. A. Kirillov for the stimulation provided by their seminars on Diophantine geometry and p-adic analysis during my stays in Moscow in 1974-75 and in Spring 1978; and to the Vietnamese mathematicians, in particular Lê-văn-Thiêm, Hà-huy-Khoái, Vương-ngọc-Châu and Đỗ-ngọc-Diệp, for their hospitality, which contributed to a fruitful and enjoyable visit to Hanoi.

Seattle
April 1980

Neal Koblitz

FRONTISPIECE: Artist's conception of the construction of the 2-adic number system as an inverse limit. By Professor A. T. Fomenko of Moscow State University.

I. BASICS

In some places in this chapter detailed proofs and computations are omitted, in order not to bore the reader before we get to the main subject matter. These details are readily available (see, for example, [53]).

1. History (very brief)

Kummer and Hensel......	1850-1900	introduced p-adic numbers and developed their basic properties
Minkowski.......	1884	proved: an equation $a_1x_1^2+\ldots+a_nx_n^2 = 0$ (a_i rational) is solvable in the rational numbers if and only if it is solvable in the reals and in the p-adic numbers for all primes p (see [13,84])
Tate............	1950	Fourier analysis on p-adic groups; pointed toward interrelations between p-adic numbers and L-functions and representation theory (see [59])
Dwork...........	1960	used p-adic analysis to prove rationality of the zeta-function of an algebraic variety defined over a finite field, part of the Weil conjectures (see [25,53])
(Kummer.........	1851	congruences for Bernoulli numbers--but he approached them in an *ad hoc* way, without p-adic numbers)
Kubota-Leopoldt	1964	interpretation of Kummer congruences for Bernoulli numbers using p-adic zeta-function
Iwasawa, Serre, Mazur, Manin, Katz, others	past 15 years	p-adic theories for many arithmetically interesting functions

Dwork, Grothendieck and their students	past 15 years	p-adic differential equations, p-adic cohomology, crystals

2. Basic concepts

Let p be a prime number, fixed once and for all. The "p-adic numbers" are all expressions of the form

$$a_m p^m + a_{m+1} p^{m+1} + a_{m+2} p^{m+2} + \dots,$$

where the $a_i \in \{0,1,2,\dots,p-1\}$ are digits, and m is any integer. These expressions form a field ($+$ and $\times$ are defined in the obvious way), which contains the nonnegative integers

$$n = a_0 + a_1 p + \dots + a_r p^r \quad (\text{"}n\text{ written to the base }p\text{"}),$$

and hence contains the field of rational numbers Q. For example,

$$-1 = (p-1) + (p-1)p + (p-1)p^2 + \dots;$$

$$\frac{-a_0}{p-1} = a_0 + a_0 p + a_0 p^2 + \dots,$$

as is readily seen by adding 1 to the first expression on the right and multiplying the second expression on the right by $1-p$.

An equivalent way to define the field Q_p of p-adic numbers is as the completion of Q under the "p-adic metric" determined by the norm $|\ |_p$: $Q \to$ nonnegative real numbers, defined by

$$\left|\frac{a}{b}\right|_p = p^{\operatorname{ord}_p b - \operatorname{ord}_p a}, \qquad |0|_p = 0,$$

where ord_p of a nonzero integer is the highest power of p dividing it. Under this norm, numbers highly divisible by p are "small", while numbers with p in the denominator are "large". For example, $|250|_5 = 1/125$, $|1/250|_5 = 125$. Clearly, $|\ |_p$ is multiplicative, because ord_p behaves like log:

$$\operatorname{ord}_p(xy) = \operatorname{ord}_p x + \operatorname{ord}_p y.$$

Also note that $|n|_p \le 1$ for n an integer.

It is not hard to verify that the completion of Q under the p-adic metric can be identified with the set Q_p of "p-adic expansions" $a_m p^m + a_{m+1} p^{m+1} + \dots$. The norm $|\ |_p$ is easy to evaluate on an element of Q_p written in its p-adic expansion: if $x = a_m p^m + a_{m+1} p^{m+1} + \ldots$ with $a_m \neq 0$, then $|x|_p = p^{-m}$.

Thus, Q_p is obtained from $|\ |_p$ in the same way as the real number field R is obtained from the usual absolute value $|\ |$: as the completion of Q. In fact, a theorem of Ostrowski (see [13] or [53]) says that any norm on Q is equivalent to the usual $|\ |$ or to $|\ |_p$ for some p. Hence, together with R, the various Q_p make up all possible completions of Q:

$$\begin{array}{ccccc} R & Q_2 & Q_3 & Q_5 \cdots & Q_p \\ & \searrow\searrow & \downarrow & \swarrow\swarrow & \\ & & Q & & \end{array}$$

Often, a situation can be studied more easily over R and Q_p than over Q; and then the information obtained can be put together to conclude something about the situation over Q. For example, one can readily show that a rational number has a square root in Q if and only if it has a square root in R and for all p has a square root in Q_p. This assertion is a special case of the Hasse-Minkowski theorem (see §1 above).

In addition to multiplicativity, the other basic property of a norm $|\ |$ on a field is the "triangle inequality" $|x + y| \leq |x| + |y|$, so named because in the case of the complex numbers C it says that in the complex plane one side of a triangle is less than or equal to the sum of the other two sides. The norm $|\ |_p$ on Q_p satisfies a stronger inequality:

$$|x + y|_p \leq \max(|x|_p, |y|_p). \qquad (2.1)$$

This is obvious if we recall how to evaluate $|x|_p$ for $x = a_m p^m + a_{m+1} p^{m+1} + \ldots$ (see above). A norm that satisfies (2.1) is called "non-Archimedean". Inequality (2.1) is sometimes called the "isosceles triangle principle", because it immediately implies that, among the three "sides" $|x|_p$, $|y|_p$ and $|x+y|_p$, at least two must be equal. Thus, in non-Archimedean geometry "all triangles are isosceles".

Here is another strange consequence of (2.1). In a field with a non-Archimedean norm $|\ |_p$, define

$$D_a(r) = \{x \mid |x-a|_p \leq r\} \text{ ("closed" disc of radius } r \text{ centered at } a)$$
$$D_a(r^-) = \{x \mid |x-a|_p < r\} \text{ ("open" disc of radius } r \text{ centered at } a). \qquad (2.2)$$

Then if $b \in D_a(r)$, it follows from (2.1) that $D_b(r) = D_a(r)$. (Also, if $b \in D_a(r^-)$, then $D_b(r^-) = D_a(r^-)$.) Thus, any point in a disc is its center! In particular, any point in a disc (or in its complement) has a neighborhood completely contained in the disc (resp., in its complement). Therefore, any disc is <u>both</u> open <u>and</u> closed in the topological sense. That is why the words "open" and "closed" in (2.2) are in quotation marks; these words are used <u>only</u> by analogy with classical geometry, and one should not be misled by them.

In Q_p, it is not hard to see that all discs of finite radius are compact. The most important such disc is

$$Z_p \underset{\text{def}}{=} D_0(1) = \{x \mid |x|_p \leq 1\} = \{x = a_0 + a_1p + a_2p^2 + \ldots\}.$$

Z_p is a ring, whose elements are called p-adic "integers". Z_p is the closure of the ordinary integers Z in Q_p. In Q_p, the other discs centered at 0 are

$$p^mZ_p = \{x = a_mp^m + a_{m+1}p^{m+1} + \ldots\} \qquad \text{for } m \in Z.$$

Z_p is a local ring, i.e., it has a unique maximal ideal pZ_p, and its residue field Z_p/pZ_p is the field of p elements $F_p = Z/pZ$. The set of invertible elements in the ring Z_p is

$$Z_p^* \underset{\text{def}}{=} Z_p - pZ_p = \{x \mid |x|_p = 1\}$$
$$= \{x = a_0 + a_1p + a_2p^2 + \ldots \mid a_0 \neq 0\}.$$

There are $p-1$ numbers in Z_p^* which play a special role: the $(p-1)$-th roots of one. For each possible choice of $a_0 = 1, 2, \ldots, p-1$, there is a unique such root whose first digit is a_0; we denote it $\omega(a_0)$ and call it the <u>Teichmüller representative</u> of a_0. For example, for $p = 5$

$\omega(1) = 1$

$\omega(2) = 2 + 1\cdot5 + 2\cdot5^2 + 1\cdot5^3 + 3\cdot5^4 + \ldots$

$\omega(3) = 3 + 3\cdot5 + 2\cdot5^2 + 3\cdot5^3 + 1\cdot5^4 + \ldots = -\omega(2)$

$\omega(4) = 4 + 4\cdot5 + 4\cdot5^2 + 4\cdot5^3 + 4\cdot5^4 + \ldots = -1.$

Except for $\omega(\pm1)$, the Teichmüller representatives are irrational, so their p-adic digits do not repeat, and can be expected to be just as random as, say, the decimal digits in $\sqrt{2}$.

If $x = a_0 + a_1p + \ldots \in Z_p^*$, we set $\omega(x) = \omega(a_0)$. Any $x \in Q_p$ can be written as $x = p^{\mathrm{ord}_p x} x_0$ for $x_0 \in Z_p^*$. Then we write

$$x = p^{\mathrm{ord}_p x}\, \omega(x_0)\, \langle x_0\rangle,$$

where $\langle x_0\rangle \underset{\mathrm{def}}{=} x_0/\omega(x_0)$ is in $1 + pZ_p$, the set of x such that $|x-1|_p < 1$.

The ring Z_p is the inverse limit of the rings Z/p^nZ with respect to the map "reduction mod p^n" from Z/p^mZ to Z/p^nZ for $m \geq n$. This suggests that, if we want to solve an equation $f(x) = 0$ for $x \in Z_p$, we should first solve it in $Z/pZ = F_p$, then in Z/p^2Z, Z/p^3Z, and so on. An important condition under which a solution in F_p can be "lifted" to a solution in Z_p is given by

Hensel's Lemma. <u>Suppose that</u> $f(x) \in Z_p[x]$, $f(a_0) \equiv 0 \pmod p$, <u>and</u> $f'(a_0) \not\equiv 0 \pmod p$ <u>(here</u> f' <u>is the formal derivative of the polynomial</u> f). <u>Then there exists a unique</u> $x = a_0 + \ldots \in Z_p$ <u>such that</u> $f(x) = 0$.

Hensel's Lemma is proved by Newton's method for approximating roots (see [59,53]).

For example, when $f(x) = x^{p-1} - 1$, any $a_0 \in \{1,\ldots,p-1\}$ satisfies $f(a_0) \equiv 0 \pmod p$, while $f'(a_0) = (p-1)a_0^{p-2} \not\equiv 0 \pmod p$; so Hensel's Lemma tells us that a_0 has a unique Teichmüller representative $\omega(a_0) \in Z_p^*$.

Unlike in the case of R, whose algebraic closure C is only a quadratic extension, Q_p has algebraic extensions of arbitrary degree; its algebraic closure $\overline{Q}_p$ has infinite degree over Q_p. Can $|\ |_p$ be extended from Q_p to $\overline{Q}_p$? Well, suppose α is algebraic over Q_p and satisfies the minimal polynomial $f(x) = x^d + a_{d-1}x^{d-1} + \ldots + a_0$. It is not hard to show that a multiplicative norm on $\overline{Q}_p$ extending $|\ |_p$ would have to be unique. So the value of this extended $|\ |_p$ on α and each of its conjugates would be the same (because we can also get an extension of $|\ |_p$ by composing our first extension of $|\ |_p$ with a field automorphism of $\overline{Q}_p$ taking α to the conjugate). Therefore, the only possible value for $|\alpha|_p$ is the d-th root of $|a_0|_p$. It turns out that this definition

$$|\alpha|_p = \sqrt[d]{|N_{Q_p(\alpha)/Q_p}(\alpha)|_p} \qquad \text{(N denotes field norm)}$$

does in fact give a norm on $\overline{Q}_p$. But the fact that this $|\ |_p$ satisfies the triangle inequality is not trivial to prove. The extension of $|\ |_p$ to $\overline{Q}_p$ is perhaps the hardest of the basic facts about p-adic numbers; for two different proofs, see [13] and [53].

We now define the ord_p function on $\overline{Q}_p$ by $\text{ord}_p\alpha = -\log_p|\alpha|_p$, so as to agree with the earlier ord_p on Q_p. (Here $\log_p$ is the ordinary "log to base p", not to be confused with a p-adic logarithm which we shall introduce shortly.) Clearly, if $[K:Q_p] = d$, then the image of K under ord_p is an additive subgroup of $\frac{1}{d}Z$, and so $\text{ord}_p K = \frac{1}{e}Z$ for some e dividing d. This positive integer e is called the <u>index of ramification</u> of K. There are two extremes:

(1) $e = 1$. Then K is called <u>unramified</u>. An example is $K = Q_p(\sqrt[N]{1})$ for N not divisible by p. In fact, it can be shown that every unramified K is contained in some cyclotomic field, so the "unramified closure" of Q_p is $Q_p^{unr} = \bigcup_{p\nmid N} Q_p(\sqrt[N]{1})$.

(2) $e = d$. Then K is called <u>totally ramified</u>. An example is $K = Q_p(\xi)$ for $\xi \neq 1$ a p-th root of one, i.e., a root of $x^{p-1} + x^{p-2} + \ldots + x + 1 = 0$. To show that K is totally rami-

fied, it suffices to find $\lambda \in K$ such that $\text{ord}_p\lambda = 1/(p-1)$. Let $\lambda = \xi-1$. Since λ satisfies: $0 = [(x+1)^p-1] / [(x+1)-1] = x^{p-1} + px^{p-2} + \frac{1}{2}p(p-1)x^{p-3} + \ldots + p$, it follows that $\text{ord}_p\lambda = \frac{1}{p-1}\,\text{ord}_p p = 1/(p-1)$. More generally, if ξ is a primitive p^n-th root of one, then $Q_p(\xi)$ is totally ramified of degree p^n-p^{n-1}, and

$$\text{ord}_p(\xi - 1) = \frac{1}{p^n - p^{n-1}}. \tag{2.3}$$

The set of all totally ramified extensions is harder to describe than the set of all unramified extensions. And, of course, "most" extensions are neither unramified nor totally ramified. In the general case we write $d = e\cdot f$.

The significance of f is as follows. If K is any field with a non-Archimedean norm $|\ |_p$, we let

$$0_K = \{x \in K|\ |x|_p \leq 1\}, \qquad M_K = \{x \in K|\ |x|_p < 1\}.$$

0_K is called the "ring of integers" of K, and M_K is the unique maximal ideal in 0_K. If K is algebraic over Q_p, then the residue field $0_K/M_K$ will be algebraic over F_p. If K has degree d and ramification index e, then this residue field has degree $f = d/e$ over F_p (see [59]).

Let us return to the case of K unramified, of degree $d = f$. Let $q = p^f$, so that $0_K/M_K$ is the field of q elements F_q. Then, using Hensel's Lemma (generalized to 0_K), we see that every nonzero element $a_0 \in F_q$ has a unique Teichmüller representative $\omega(a_0) \in K$ such that $\omega(a_0)^{q-1} = 1$ and $\omega(a_0)$ mod M_K is a_0. If a_0 generates F_q as an extension of F_p, then $K = Q_p(\omega(a_0))$. These Teichmüller representatives are a natural choice of "digits" in K: every $x \in K$ can be written uniquely as the limit of a sum

$$x = \Sigma_{i\geq m}\, a_i p^i, \quad \text{where } a_i \in \{\omega(a)\}_{a\in F_q}$$

(we agree to let $\omega(0) = 0$). Even in Q_p it is sometimes convenient to choose $0, \omega(1), \omega(2),\ldots, \omega(p-1)$ as digits instead of $0, 1, 2,\ldots, p-1$.

Since the complex number field C is a finite dimensional R-vector space, it is complete under the extension of $| \ |$ to C. However, $\overline{Q}_p$ turns out not to be complete under $| \ |_p$. For example, the convergent infinite sum $\Sigma x_i p^i$, where the x_i are a sequence of roots of one of increasing degree, in general is not algebraic over Q_p. Thus, in order to do analysis, we must take a larger field than $\overline{Q}_p$. We denote the completion of $\overline{Q}_p$ by Ω_p:

$$\Omega_p = \hat{\overline{Q}}_p \quad (\hat{} \text{ means completion with respect to } | \ |_p).$$

It is not hard to see that Ω_p is algebraically closed, as well as complete, that $0_{\Omega_p}/M_{\Omega_p} = \overline{F}_p$, and that $\text{ord}_p \Omega_p = Q$. Sometimes Ω_p is denoted C_p in order to emphasize the analogy with the complex numbers (i.e., both are the smallest extension field of Q that is both algebraically closed and complete in the respective metric). But in some respects Ω_p is more complicated. For example, it is a much bigger extension of Q_p than C is over R (in fact, Ω_p has uncountable transcendance degree over Q_p), and it is easy to see that Ω_p is <u>not</u> locally compact.

3. Power series

An infinite sum Σa_i has a limit if $\Sigma_{N\leq i<M} a_i$ is small for large N, $M > N$. Because of the isosceles triangle principle (2.1), in Ω_p this occurs if and only if $a_i \to 0$, i.e., $|a_i|_p \to 0$, or equivalently, $\text{ord}_p a_i \to \infty$. Thus, the question of convergence or divergence of a power series $\Sigma a_i x^i$ depends only on $|x|_p$, not on the precise value of x. There is no "conditional convergence". Thus, every infinite series $\Sigma a_i x^i$ has a radius of convergence r such that one of the following holds:

$$\sum_{i=0}^{\infty} a_i x^i \text{ converges} \iff x \in D(r^-) \quad (\underset{\text{def}}{=} D_0(r^-), \text{ see (2.2)})$$

or

$$\sum_{i=0}^{\infty} a_i x^i \text{ converges} \iff x \in D(r) \quad (\underset{\text{def}}{=} D_0(r)).$$

An example of the first alternative is Σx^i (where $r = 1$); an example of the second is the derivative $\Sigma p^i x^{p^i - 1}$ of Σx^{p^i} (here also $r = 1$).

An important example is the series $e^x = \Sigma x^i/i!$. To determine its radius of convergence, we must find $\text{ord}_p(i!)$. If i is a power of p, it is easy to see that $\text{ord}_p(p^n!) = p^{n-1} + p^{n-2} + \ldots + p + 1 = (i-1)/(p-1)$. More generally, if we write the positive integer i to the base p: $i = \Sigma a_i p^i$, and let $S_i = \Sigma a_i$ denote the sum of its digits, then

$$\text{ord}_p(i!) = \frac{i - S_i}{p - 1}. \tag{3.1}$$

Since $1 \leq S_i \leq (p-1)(\log_p i + 1)$, it follows that asymptotically $\text{ord}_p(i!) \sim \frac{i}{p-1}$, and so

$$\text{ord}_p(x^i/i!) \longrightarrow \infty \iff \text{ord}_p x > \frac{1}{p-1}$$

$$\iff x \in D(\tfrac{1}{\gamma}^-), \quad \gamma = \sqrt[p-1]{p} > 1.$$

Thus, e^x converges in a disc smaller than the unit disc. In the classical case the $i!$ in the denominator makes e^x converge everywhere, but in $|\ |_p$ it has a harmful effect on convergence. The poor convergence of e^x causes much of p-adic analysis, e.g., differential equations, to involve subtleties which are absent in complex analysis.

To obtain a series convergent in $D(1^-)$ instead of $D(\frac{1}{\gamma}^-)$, we can replace e^x by $e^{\pi x}$, where π (not to be confused with the real number $\pi = 3.14\ldots$) is any element of Ω_p such that $\text{ord}_p \pi = 1/(p-1)$. The best choice of π is a $(p-1)$-th root of $-p$, for reasons that will become clear later.

We can analyze more closely why e^x converges so poorly if we use the formal power series identity

$$e^x = \prod_{n=1}^{\infty} (1 - x^n)^{-\mu(n)/n} \quad \text{in } \mathbb{Q}[[x]], \tag{3.2}$$

where the Möbius function μ is defined by

$$\mu(n) = \begin{cases} 0 & \text{if there is a prime whose square divides } n; \\ (-1)^k & \text{if } n \text{ is a product of } k \text{ distinct primes.} \end{cases}$$

The identity (3.2) is easily proved by taking log of both sides and using the fact that $\Sigma_{d|n}\,\mu(d) = 1$ if $n = 1$ and 0 otherwise.

Most of the terms in (3.2) -- those for which $p\nmid n$ -- have fairly good convergence, because the binomial series

$$(1 + Y)^\alpha = \sum \binom{\alpha}{i} Y^i, \quad \binom{\alpha}{i} = \frac{\alpha(\alpha-1)\cdots(\alpha-i+1)}{i!}$$

has coefficients

$$\binom{\alpha}{i} \in Z_p \quad \text{for} \quad \alpha \in Z_p. \tag{3.3}$$

(Namely, this is trivial for α a positive integer; then use the fact that the positive integers are dense in Z_p.) Thus, for $p\nmid n$, $\prod(1-x^n)^{-\mu(n)/n} \in Z_p[[x]]$ and so converges for $|x|_p < 1$. The bad convergence of (3.2) comes from those n which are divisible by p. So, to get better convergence, we can define the "Artin-Hasse exponential"

$$E_p(x) = \prod_{p\nmid n} (1-x^n)^{-\mu(n)/n} = e^{x + x^p/p + x^{p^2}/p^2 + x^{p^3}/p^3 + \cdots},$$

where the last equality of formal power series is proved in the same way as (3.2). Then $E_p(x)$ is in $Z_p[[x]]$, and so converges in $D(1^-)$.

If we make the change of variables $E_p(\pi x)$, where $\pi^{p-1} = -p$, the first two terms in the exponent are $\pi(x - x^p)$. The expression $x - x^p$ plays a key role in much of p-adic analysis, since in a field of characteristic p

$$x \in \text{the prime field } F_p \Leftrightarrow x - x^p = 0;$$

also recall that the Teichmüller representatives $\{\omega(a)\}_{a\in F_p}$ are solutions of this equation in Q_p. π is chosen to be a (p-1)-th root of $-p$ precisely so that the first two terms in the exponent for $E_p(\pi x)$ become a multiple of $x - x^p$.

Since

$$e^{\pi(x-x^p)} = E_p(\pi x) \prod_{i\geq 2} e^{-(\pi x)^{p^i}/p^i}, \qquad (3.4)$$

the convergence of $e^{\pi(x-x^p)}$ is determined by the worst convergence that occurs on the right. $E_p(\pi x)$ converges on $D(\gamma^-)$ (recall $\gamma = p^{1/(p-1)}$), and it is easy to compute that the worst series is the first one in the product, $\exp(-\pi^{p^2} x^{p^2}/p^2)$, which converges for $\text{ord}_p x > -(p-1)/p^2$. Thus, if we let $\gamma_1 = p^{(p-1)/p^2} > 1$, it follows that $e^{\pi(x-x^p)}$ converges on $D(\gamma_1^-)$, a disc strictly bigger than $D(1)$.

Thus, the $-x^p$ in $e^{\pi(x-x^p)}$ is a "correction" which improves the convergence of $e^{\pi x}$. We can see how this works if we look at the expansions of $e^{\pi x}$ and $e^{\pi(x-x^p)}$ out to the x^p-term, the first term where the two series differ. In the expansion $e^{\pi x} = \Sigma\ (\pi x)^i/i!$, the x^p-term is the first one in which $1/i! \notin \mathbb{Z}_p$, i.e., the first term containing a p in the denominator. Thus, $|\pi^p/p!|_p = |-\pi/(p-1)!|_p = |\pi|_p$ (the first equality because $\pi^{p-1} = -p$). But the coefficient of x^p in $e^{\pi(x-x^p)}$ is

$$\frac{\pi^p}{p!} - \pi = \pi(-p/p! - 1) = \frac{-\pi}{(p-1)!}(1 + (p-1)!).$$

A simple fact of elementary number theory (Wilson's theorem) says: $(p-1)! \equiv -1 \pmod p$. Hence, the p-adic norm of the coefficient of x^p in $e^{\pi(x-x^p)}$ is bounded by $|p\pi|_p = |\pi^p|_p$. Thus, the correction term $-\pi x^p$ has the effect of canceling the p in the denominator of $(\pi x)^p/p!$.

We denote $E_\pi(x) = e^{\pi(x-x^p)}$ (not to be confused with $E_p(x)$). Note that $E_\pi(x)$ must <u>first</u> be expanded as a power series and <u>then</u> evaluated. If $|x|_p < 1$, the result will be the same as if we first substituted x in $\pi(x-x^p)$ and then took the exponential. But if $|x|_p \geq 1$, that exponential will not converge unless $|x-x^p|_p < 1$, and even in the latter case will in general give the wrong value; for example, $E_\pi(1) \neq 1 = e^0$ (see §III.5).

Another important series is

$$\log(1 + x) = \sum_{i=1}^{\infty} \frac{(-1)^{i+1}}{i} x^i, \tag{3.5}$$

which is easily seen to converge on $D(1^-)$. It has better convergence than e^x! Since the identity

$$\log(xy) = \log x + \log y \tag{3.6}$$

holds as a formal power series identity, i.e., $\Sigma(-1)^{i+1}x^i/i + \Sigma(-1)^{i+1}y^i/i = \Sigma(-1)^{i+1}(x+y+xy)^i/i$ in $Q[[x,y]]$, it follows that (3.6) holds in Ω_p as long as $|x-1|_p < 1$ and $|y-1|_p < 1$. In particular, since $|\xi-1|_p < 1$ for ξ any p^n-th root of one (see §2), we can apply (3.6) to conclude that $\log \xi = 0$.

The p-adic logarithm has a natural extension to $\Omega_p^* = \Omega_p-\{0\}$, which we shall denote $\ln_p$ (so as not to confuse it with the classical log-to-the-base-p; however, in the literature $\log_p$ is normally used rather than $\ln_p$).

Proposition. <u>There exists a unique function</u> $\ln_p : \Omega_p^* \longrightarrow \Omega_p$ <u>such that</u>

(1) $\ln_p(1 + x)$ <u>is given by the series (3.5) if</u> $|x|_p < 1$;

(2) (3.6) <u>holds for all</u> $x, y \in \Omega_p^*$;

(3) $\ln_p(p) = 0$.

The third condition is a normalization, which is necessary because, as mentioned before, ord_p behaves like a logarithm. Thus, if $\ln_p$ is any function satisfying (1) and (2), then for any constant $c \in \Omega_p$ the function $\ln_p + c\cdot\text{ord}_p$ also satisfies (1) and (2).

I won't prove this proposition, but will discuss concretely how one computes a logarithm. First, for every $\frac{m}{n} \in Q$, choose "$p^{m/n}$" to be any root of $x^n - p^m = 0$. Now suppose we want to find $\ln_p x$ for some nonzero $x \in \Omega_p$. First write $x = p^{m/n}x_0$, where $m/n = \text{ord}_p x$. Since $|x_0|_p = 1$, its reduction modulo M_{Ω_p} is a nonzero element $\overline{x}_0 \in \overline{F}_p$. Let $\omega(\overline{x}_0)$ be the Teichmüller representative of $\overline{x}_0$. Then

$$x = p^{\mathrm{ord}_p x}\, \omega(\bar{x}_0)\, \langle x_0 \rangle, \quad \text{where} \quad |\langle x_0 \rangle - 1|_p < 1.$$

Since $\ln_p p = 0$, and (3.6) implies that $\ln_p(\text{any root of } 1) = 0$, we have

$$\ln_p x = \ln_p \langle x_0 \rangle = \sum (-1)^{i+1} (\langle x_0 \rangle - 1)^i / i.$$

For example,

$$\ln_5\left(\frac{1}{250}\right) = \ln_5\left(\frac{2 + 1\cdot 5 + 2\cdot 5^2 + 1\cdot 5^3 + 3\cdot 5^4 + \ldots}{2}\right)$$

$$= \sum (-1)^{i+1} \, (3\cdot 5 + 3\cdot 5^2 + 0\cdot 5^3 + 4\cdot 5^4 + \ldots)^i / i.$$

Note that a function such as ord_p, which is locally constant on Ω_p^* (i.e., for every $a \in \Omega_p^*$ there exists r such that $\mathrm{ord}_p x = \mathrm{ord}_p a$ for $x \in D_a(r)$) but is not constant, could not exist on C^*. For this reason, the theory of analytic continuation is more complicated on Ω_p^*. Unlike the classical log, $\ln_p$ is not obtained by "analytic continuation" of the series (3.5); any of the functions $\ln_p + c\cdot \mathrm{ord}_p$ would also be locally analytic and agree with (3.5) on $D_1(1^-)$.

There is a notion of p-adic global analyticity, due to Krasner [57], such that two globally analytic functions which agree, say, on a disc, must agree everywhere. Namely, let $D \subset \Omega_p$ be a so-called "quasi-connected" set, the most important examples of which are discs from which finitely many smaller discs and/or compact subsets have been removed. Then a function $f\colon D \to \Omega_p$ is said to be Krasner analytic if D is a union of open sets D_i, $D_i \subset D_{i+1}$, such that for each i, $f|_{D_i}$ is a uniform limit of rational functions having no poles in D_i. $\ln_p$ is not Krasner analytic on Ω_p^*. Later we shall see examples of interesting Krasner analytic functions. For example, the second derivative of the p-adic log gamma function turns out to be Krasner analytic on the complement of Z_p: $D = \Omega_p - Z_p$ (see p. 134).

A final remark about $\ln_p$: it has the expected derivative $\frac{1}{x}$, since $\lim_{\varepsilon \to 0} [(\ln_p(x + \varepsilon) - \ln_p x)/\varepsilon] = \lim \frac{1}{\varepsilon} \ln_p(1 + \frac{\varepsilon}{x})$, and $\ln_p(1 + \frac{\varepsilon}{x})$ is given by the usual series as soon as $|\varepsilon|_p < |x|_p$.

4. Newton polygons

a. Classical case

For $f(X,Y) = \Sigma a_{ij}X^iY^j \in R[X,Y]$, let M_f be the convex hull of the following set of points in the (i,j)-plane: $\{(i,j) \mid a_{ij} \neq 0\}$. M_f is called the Newton polygon of f. If two polynomials f, g $\in R[X,Y]$ have no common factors, then the two curves determined by f and g intersect in a finite number N of points (counting multiplicity): $\{(x,y) \mid f(x,y) = g(x,y) = 0\}$. Let $M_f + M_g = \{z = x + y \mid x \in M_f,\ y \in M_g\}$. Then it can be shown that

$$N \leq \text{area}(M_f + M_g) - \text{area}(M_f) - \text{area}(M_g).$$

b. The p-adic case: polynomials

Let $f(x) = a_0 + \ldots + a_dx^d \in \Omega_p[x]$. The Newton polygon M_f of f is defined to be the convex hull of the points (i, ord_pa_i) (where we agree to take $\text{ord}_p0 = +\infty$), i.e., M_f is the polygonal line obtained by rotating a vertical line through $(0, \text{ord}_pa_0)$ counterclockwise until it bends around various points (i, ord_pa_i), and eventually reaches the point (d, ord_pa_d). This is similar to the classical case (where we take $a_i = \Sigma_j a_{ij}Y^j$ and ord_Ya_i = the least j for which $a_{ij} \neq 0$), except that we only take the lower part of the convex hull.

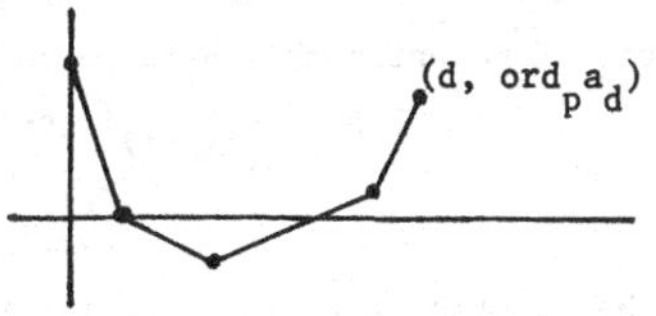

It is not hard to prove the following

Proposition. _If a segment of_ M_f _has slope_ λ _and horizontal length_ N _(i.e., it extends from_ (i, ord_pa_i) _to_ $(i+N, \lambda N + \text{ord}_pa_i))$, _then_ f _has precisely_ N _roots_ r_i _with_ $\text{ord}_pr_i = -\lambda$ _(counting multiplicity)_.

Examples. (1) The Eisenstein irreducibility criterion: if $f(x) = a_0 + \ldots + a_{d-1}x^{d-1} + x^d \in Q[x]$, and if there exists a prime p such that $\text{ord}_pa_i \geq 1$ for $0 \leq i < d$ and $\text{ord}_pa_0 = 1$, then f is irreducible over Q. In fact, using the Newton polygon M_f, we can quickly see that f is even irreducible over Q_p. Namely, the conditions on $\text{ord}_p a_i$ imply that M_f consists of the line segment from $(0,1)$ to $(d,0)$. Hence f has d roots all of ordinal $\frac{1}{d}$. If f factored over Q_p, each root r would have degree $d' < d$ over Q_p, and hence we would have $\text{ord}_p r \in \frac{1}{d'}Z$. Thus, f is irreducible.

(2) Later we'll want to study the curve $y^p - y = x^d$. If this curve is considered over a field of characteristic p, there are p obvious automorphisms $x \mapsto x$, $y \mapsto y+\bar{a}$, $\bar{a} \in F_p$. Suppose we want to find similar automorphisms $x \mapsto x$, $y \mapsto y+a$ when the curve is considered over Ω_p. For example, let us fix $y \in \Omega_p$ and look for $a \in \Omega_p$ such that sending $y \mapsto y+a$ "lifts" the automorphism $y \mapsto y+1$ in the sense that $a = 1 + z$ with $|z|_p < 1$, i.e., $a \equiv 1 \pmod{M_{\Omega_p}}$. It is convenient to suppose that $|y|_p < \gamma$, where $\gamma = p^{1/(p-1)} > 1$. We must choose z so that

$$(y + 1 + z)^p - (y + 1 + z) = y^p - y,$$

or, if we write this as a polynomial in z,

$$z^p + \sum_{i=1}^{p-2} \binom{p}{i} (y+1)^i z^{p-i} + (p(y+1)^{p-1}-1)z + $$
$$+ [(y+1)^p - y^p - 1] = 0.$$

The constant term $a_0 = (y+1)^p - y^p - 1 = \Sigma_{1 \le i < p} \binom{p}{i} y^i$ satisfies $\text{ord}_p a_0 \geq 1 + \min(0, (p-1)\text{ord}_p y)$, which is greater than zero, since we have assumed that $\text{ord}_p y > -1/(p-1)$. On the other hand, $\text{ord}_p a_1 = \text{ord}_p a_p = 0$ and $\text{ord}_p a_i \geq 0$ for $1 < i < p$. Hence, the Newton polygon of this polynomial in z is as shown in the diagram to the right. The only nonzero slope is the first little segment, with slope $\lambda = -\text{ord}_p a_0$. Thus, there is exactly one root z with $|z|_p < 1$, in fact, with $\text{ord}_p z = -\lambda = \text{ord}_p a_0$. This root z gives the unique lifting to Ω_p of the automorphism $y \mapsto y+1$ in characteristic p. The other $p-1$ roots z have $|z|_p = 1$, and the corresponding maps $y \mapsto y+1+z$ lift the other automorphisms $y \mapsto y+\bar{a}$, $\bar{a} \in F_p$.

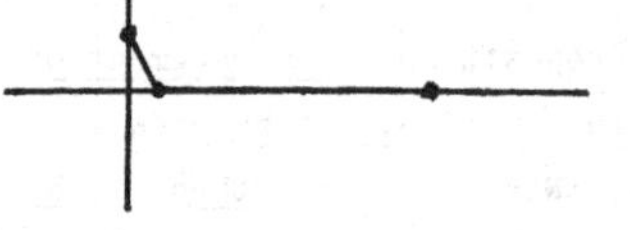

c. The p-adic case: power series

The Newton polygon M_f for a power series $f(x) = \Sigma a_i x^i \in \Omega_p[[x]]$ is defined just as for polynomials, except that now it extends infinitely far to the right. Also, it is possible for the Newton polygon to include an infinitely long segment without any points $(i, \text{ord}_p a_i)$ far to the right. For example, the power series $1 + \Sigma_{j\geq 1} p^{j-1}x^{p^j}$ has simply the x-semiaxis as its Newton polygon, although $\text{ord}_p a_i > 0$ for $i > p$. Here is the case $p = 2$:

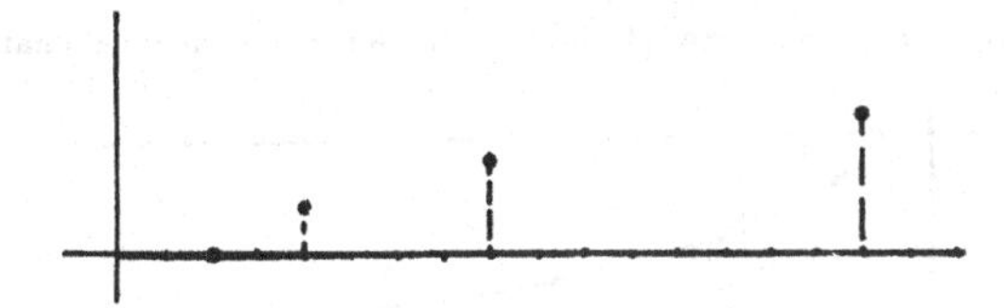

The following theorem is the p-adic analog of the Weierstrass Preparation Theorem.

Theorem. *Let* $f(x) = a_m x^m + \ldots \in \Omega_p[[x]]$, $a_m \neq 0$, *be a power series which converges on* $D(p^\lambda)$. *Let* $(N, \text{ord}_p a_N)$ *be the right endpoint of the last segment of* M_f *with slope* $\leq \lambda$, *if this* N *is finite. Otherwise, there will be a last infinitely long segment of slope* λ *and only finitely many points* $(i, \text{ord}_p a_i)$ *on that segment. In that case let* N *be the last such* i *(for example, in the above illustration* $N = 2$*). Then there exists a unique polynomial* $h(x)$ *of the form* $b_m x^m + b_{m+1}x^{m+1} + \ldots + b_N x^N$ *with* $b_m = a_m$ *and a unique power series* $g(x)$ *which converges and is nonzero on* $D(p^\lambda)$, *such that*

$$f(x) = \frac{h(x)}{g(x)} \quad \text{on } D(p^\lambda).$$

In addition, M_h *coincides with* M_f *as far as the point* $(N, \text{ord}_p a_N)$.

Corollary 1. *Within the region of convergence of* f, *the Newton polygon determines* ord_p *of the zeros of* f *in the same way* as for polynomials.

Corollary 2. A power series which converges everywhere and has no zeros is a constant.

For proofs of these facts, see, for example, [53].

Examples. (1) The power series $1 + \Sigma_{j\geq 1} p^{j-1}x^{p^j}$, which converges on $D(1)$, has precisely p zeros, all with $|\ |_p = 1$.

(2) For the log series $f(x) = \Sigma\ (-1)^{i+1}\ x^i/i$, M_f is the polygonal line connecting the points $(p^j,-j)$, $j = 0, 1, 2, \ldots$. The picture for $p = 2$ is given below. We may conclude that in

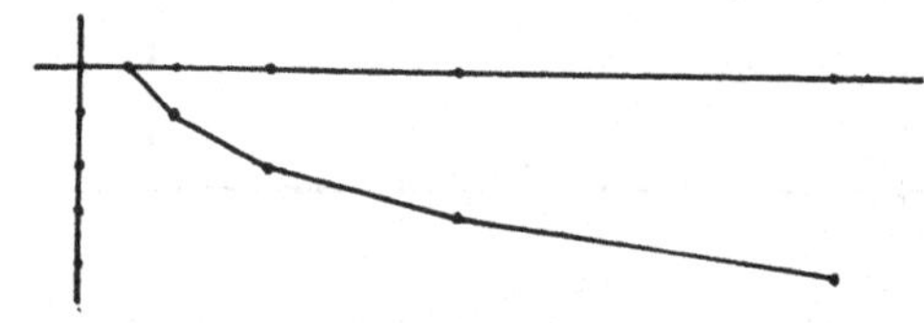

$D_1(1^-)$ the function $\ln_p$ vanishes at points $1 + x$ for exactly $p^j - p^{j-1}$ values of x with ordinal $1/(p^j - p^{j-1})$. These x's are precisely $x = \xi - 1$ for ξ a primitive p^j-th root of one (see (2.3)).

Remark. Some specialists prefer another definition of the Newton polygon. Instead of the points $(i, \operatorname{ord}_p a_i)$, they look at the lines ℓ_i: $y = ix + \operatorname{ord}_p a_i$ with slope i and y-intercept $\operatorname{ord}_p a_i$. Then $\tilde{M}_f$ is defined as the graph of the function $\min_i \ell_i(x)$. The x-coordinates of the points of intersection of the ℓ_i which appear in $\tilde{M}_f$ give ord_p of the zeros, and the difference between the slopes i of successive ℓ_i which appear in $\tilde{M}_f$ give the number of zeros with given ord_p. For example, $\tilde{M}_f$ for the log series $f(x) = \Sigma\ (-1)^{i+1}\ x^i/i$ is shown in the drawing on the next page. It somewhat resembles the usual graph of log, especially near the y-axis. This type of Newton polygon was used in Hà-huy-Khoái's thesis [39], which contains a detailed discussion of such Newton polygons, as well as a new generalization of Newton polygons

("Newton sequences") which can be used for more refined investigations of power series.

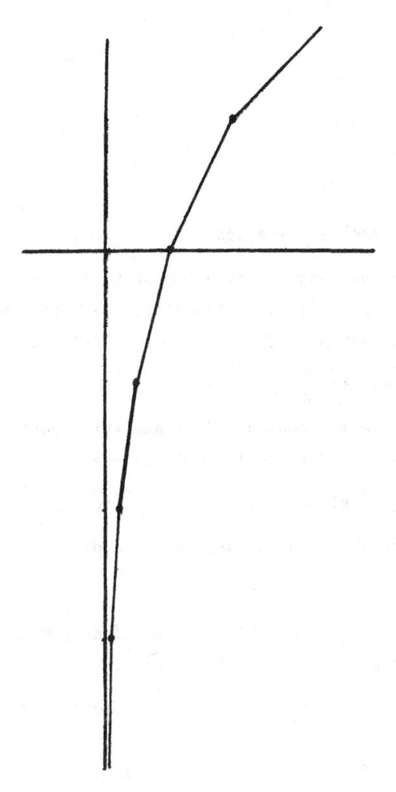

II. p-ADIC ζ-FUNCTIONS, L-FUNCTIONS, AND Γ-FUNCTIONS

1. Dirichlet L-series

We leave p-adics for a moment to review the basic facts about Dirichlet L-series (see [13,41]). Let $f\colon Z \to C$ be a periodic function with period d: $f(x+d) = f(x)$. Then we define

$$L(s,f) = \sum_{n=1}^{\infty} f(n)\, n^{-s}$$

for $\operatorname{Re} s > 1$, and extend by analytic continuation to other $s \in C$. The generalized Bernoulli numbers are

$$B_{k,f} = k!\cdot\text{coefficient of } t^k \text{ in } \sum_{a=0}^{d-1} \frac{f(a)te^{at}}{e^{dt}-1}. \tag{1.1}$$

It can be shown [41] that for k a positive integer

$$L(1-k,f) = -\frac{B_{k,f}}{k}. \tag{1.2}$$

For example, for the Riemann zeta function $\zeta(s) = L(s,1)$ (where 1 denotes the constant function 1, having period 1)

$$\zeta(1-k) = -\frac{1}{k} B_k, \quad B_k = k!\cdot\text{coefficient of } t^k \text{ in } \frac{t}{e^t - 1}.$$

When $f = \chi$ is a character, i.e., a homomorphism $\chi\colon (Z/dZ)^* \to C^*$ from the multiplicative group of integers mod d (where χ is extended by $\chi(n) = 0$ for all n having a common factor with d), the L-series equals the following "Euler product" if $\operatorname{Re} s > 1$:

$$L(s,\chi) = \prod \left(1 - \frac{\chi(\ell)}{\ell^s}\right)^{-1}, \tag{1.3}$$

where the product is taken over all primes ℓ.

L-functions occur in many situations in number theory. To give a simple example, the class number h of an imaginary quadratic field $Q(\sqrt{-d})$ of discriminant $-d$ is given by

$$h = \frac{w\sqrt{d}}{2\pi} L(1,\chi) = -\frac{w}{2}\frac{1}{d}\sum_{a=1}^{d-1} a\,\chi(a),$$

where $w = 2, 4,$ or 6 is the number of roots of unity in $Q(\sqrt{-d})$, and $\chi: (Z/dZ)^* \longrightarrow \{\pm 1\}$ is the Legendre symbol (quadratic residue symbol). By the way, no elementary proof (not using the Dirichlet formula $h = -\frac{1}{d}\Sigma\, a\,\chi(a)$) is known for the nonvanishing of the simple sum $\Sigma\, a\,\chi(a)$. Later (Chapter IV) we shall study generalizations and p-adic analogs of the formula $L(1,\chi) = 2\pi h/w\sqrt{d}$.

We shall also want to consider "twisted" L-functions. Let r be a positive integer, and let $\varepsilon \neq 1$ be any nontrivial r-th root of one. Let $z^d = \varepsilon$. Then let

$$L(s,f,z) = \sum_{n=1}^{\infty} f(n)\, z^n\, n^{-s}.$$

Since the function $n \mapsto f(n)z^n$ has period dr, this is a special case of the L-series considered above. In particular, if we replace k by $k+1$ and $f(a)$ by $z^a f(a)$ in (1.1) and (1.2), we obtain:

$$L(-k,f,z) = k!\cdot\text{coefficient of } t^k \text{ in } \sum_{\substack{0\le a<d\\ 0\le b<r}} \frac{f(a)e^{at}e^{bdt}z^{a+bd}}{1-e^{rdt}}$$

$$= k!\cdot\text{coefficient of } t^k \text{ in } \sum_{0\le a<d} \frac{f(a)z^a e^{at}}{1-\varepsilon e^{dt}}. \qquad (1.4)$$

We now proceed to the p-adic theory.

2. p-adic measures

Let d be a fixed positive integer, and let $X = \varprojlim_N Z/dp^N Z$, where the map from $Z/dp^M Z$ to $Z/dp^N Z$ for $M \ge N$ is reduction mod dp^N. In the special case $d = 1$, X is simply Z_p. By $a + dp^N Z_p$ we mean the set of all $x \in X$ which map to a under the natural map $X \longrightarrow Z/dp^N Z$. Without loss of generality, we may agree always to choose a so that $0 \le a < dp^N$. Note that

$X = \bigsqcup_{0\leq a<d} a + dZ_p$ is a disjoint union of d topological spaces isomorphic to Z_p. Also,

$$a + dp^N Z_p = \bigsqcup_{0\leq b<p} (a+bdp^N) + dp^{N+1} Z_p \qquad \text{(disjoint union)}. \quad (2.1)$$

It is not hard to show that any open subset which is compact (i.e., closed, since X is compact) is a finite union of compact-open sets of the form $a + dp^N Z_p$. (Warning: Not all open sets are compact, for example, $X-\{0\}$.)

Definition. An Ω_p-valued measure μ on X is a finitely additive bounded map from the set of compact open $U \subset X$ to Ω_p.

If we are given the values of a function μ only on the sets $a + dp^N Z_p$, such a μ extends to a measure on all compact-open U if and only if these values are bounded and for all a

$$\mu(a + dp^N Z_p) = \sum_{b=0}^{p-1} \mu((a+bdp^N) + dp^{N+1} Z_p), \qquad (2.2)$$

i.e., we need only check additivity for the disjoint unions (2.1).

An equivalent definition of a measure is: a bounded linear functional $f \longmapsto \int f d\mu$ on the Ω_p-vector space of locally constant functions on X (i.e., functions which are a finite linear combination of characteristic functions of compact-open sets).

A routine verification shows that, if $f: X \longrightarrow \Omega_p$ is any continuous function, and we write f as a uniform limit of locally constant functions f_i, then the limit of the Riemann sums $\int f_i d\mu$ exists and depends only on f: $\int f d\mu = \lim \int f_i d\mu$. For example, we can evaluate $\int f d\mu$ as the limit

$$\int f d\mu = \lim_{N\to\infty} \sum_{0\leq a<dp^N} f(a)\, \mu(a + dp^N Z_p). \qquad (2.3)$$

Clearly, the Ω_p-valued measures form an Ω_p-vector space. For more detailed proofs, see, for example, [53].

Remark. Much more general p-adic measures have been defined: measures on more general types of p-adic spaces X (Mazur, Manin,

Katz), measures which take values in spaces of modular forms (Katz) or spaces of operators (Vishik), unbounded measures (Manin, Vishik).

Basic example. Fix $z \in \Omega_p$ so that $\varepsilon = z^d$ is <u>not</u> in $D_1(1^-)$. Then $|\varepsilon^{p^N}-1|_p \geq 1$ for all N. The most important case is when $\varepsilon = z^d$ is a root of one which is not a p^N-th root of one for any N. Define

$$\mu_z(a + dp^N Z_p) = \frac{z^a}{1 - \varepsilon^{p^N}}.$$

This gives a measure, since boundedness is ensured by stipulating that ε^{p^N} is not close to 1, and the verification of additivity reduces to summing a geometric progression:

$$\sum_{b=0}^{p-1} \mu_z(a+bdp^N + dp^{N+1}Z_p) = \frac{1}{1 - \varepsilon^{p^{N+1}}} \sum_{b=0}^{p-1} z^{a+bdp^N}$$

$$= \frac{z^a}{1 - \varepsilon^{p^{N+1}}} \sum_{b=0}^{p-1} \varepsilon^{bp^N} = \frac{z^a}{1 - \varepsilon^{p^N}} = \mu_z(a + dp^N Z_p).$$

An especially simple case, considered by Osipov [78], occurs when $d = 1$ and $z = \varepsilon$ is a $(p-1)$-th root of one, in which case the denominator $1 - \varepsilon^{p^N} = 1 - \varepsilon$ is simply a constant.

Since the space X "brings together" Z/dZ and Z_p, we have two natural sources of continuous functions on X. (1) Any $f: Z \longrightarrow \Omega_p$ having period d can be considered as a continuous (in fact, locally constant) function on X by setting $f(x) = f(a)$ for $x \in a + dZ_p$. (2) Any continuous $f: Z_p \longrightarrow \Omega_p$ can be pulled back to X by means of the map from X to Z_p which "forgets mod d information" (i.e., the map which is the inverse limit of the projections reduction mod p^N: $Z/dp^N Z \longrightarrow Z/p^N Z$).

We shall look at the following example of the second type of continuous function on X. Let $t \in \Omega_p$ be any small fixed value (namely, $\mathrm{ord}_p t > 1/(p-1)$). Then $e^{tx} = \Sigma\, t^i x^i/i!$ is a continuous function on X; its value at an $x \in X$ is determined by approximating x by a for which $x \in a + dp^N Z_p$.

Now let f be a function of period d, and consider the function $e^{tx}f(x)$ on X. We can integrate this function using (2.3) and summing the geometric progression. We shall write $d\mu_z(x)$ to remind ourselves that x (and not t) is the variable of integration. We have:

$$\int e^{tx}f(x)\, d\mu_z(x) = \lim_{N\to\infty} \frac{1}{1-\varepsilon^{p^N}} \sum_{0\le a<dp^N} e^{at}f(a)\, z^a$$

$$= \sum_{a=0}^{d-1} f(a)z^a e^{at} \lim_{N\to\infty} \frac{1}{1-\varepsilon^{p^N}} \sum_{b=0}^{p^N-1} (ze^t)^{bd}$$

$$= \sum_{a=0}^{d-1} \frac{f(a)z^a e^{at}}{1-\varepsilon e^{dt}} \lim_{N\to\infty} \frac{1-\varepsilon^{p^N} e^{dp^N t}}{1-\varepsilon^{p^N}}.$$

Since $e^{dp^N t}$ approaches 1 as $N\to\infty$, the limit is 1, and we obtain

$$\int e^{tx}f(x)\, d\mu_z(x) = \sum_{a=0}^{d-1} \frac{f(a)z^a e^{at}}{1-\varepsilon e^{dt}}. \tag{2.4}$$

Notice that the right side of (2.4) is the same function that appeared in the expression for $L(-k,f,z)$ in §1, except that in (1.4) the values of z and f were complex, while in (2.4) they are p-adic. The most important case of (1.4) occurs when f takes algebraic values, for example, when $f = \chi\colon (Z/dZ)^* \to C^*$ is a character. Thus, suppose that in (1.4) both z and the values of f are contained in a finite extension K of Q. If we imbed K in Ω_p, we can identify z and $f(a)$ simultaneously as complex or as p-adic numbers.

To construct such an imbedding, choose any prime ideal P of K dividing p. Introduce the "P-adic" topology on K in the same way as the p-adic topology was introduced on Q: $x \in K$ is considered to be small if the fractional ideal (x) is divisible by a large positive power of P. Then complete K in this topology. Since $P|(p)$, the resulting complete field K_P contains Q_p, and is an algebraic extension of Q_p. For more details, see [59]. In what follows, we shall suppose that such an imbedding $\iota_p: K \hookrightarrow \Omega_p$ has been chosen once and for all, so that any expression involving complex numbers which all lie in K can be simultaneously viewed as a p-adic expression.

In particular, $L(-k,f,z)$ can be considered p-adically. Then, comparing (2.4) with (1.4), we obtain

$$\sum_{k=0}^{\infty} L(-k,f,z) \frac{t^k}{k!} = \int e^{tx} f(x)\, d\mu_z(x)$$
$$= \sum_{k=0}^{\infty} \int x^k f(x)\, d\mu_z(x) \frac{t^k}{k!}.$$

Since this holds for all t with $\mathrm{ord}_p t > 1/(p-1)$, we can equate coefficients and obtain

$$L(-k,f,z) = \int x^k f(x)\, d\mu_z(x). \tag{2.5}$$

As an application of (2.5), one can now study p-adically the values at $-k$ of the Riemann zeta function, since, if we take any positive integer r prime to p, we have

$$\sum_{\varepsilon^r=1,\ \varepsilon\neq 1} L(s,1,\varepsilon) = \sum_{n=1}^{\infty} n^{-s} \begin{cases} r-1 & \text{if } r\mid n \\ -1 & \text{if } r\nmid n \end{cases} = (r^{1-s}-1)\,\zeta(s).$$

Thus, for $d = 1$, $X = Z_p$, and μ defined as the sum of μ_ε over all ε with $\varepsilon^r = 1$, $\varepsilon \neq 1$, we have

$$\zeta(-k) = \frac{1}{r^{k+1}-1} \int x^k\, d\mu(x). \tag{2.6}$$

Remark. The relation between this μ and Mazur's measures μ_α (see [53]) is that $\mu = \mu_{\mathrm{Mazur},\alpha}$ for $\alpha = 1/r$.

3. p-adic interpolation

For simplicity, we first treat the case of the Riemann zeta function, and take $d = 1$, $X = Z_p$. We know that the values

$$\zeta(1-k) = -\frac{1}{k} B_k = \frac{1}{r^k-1} \int x^{k-1}\, d\mu(x) \tag{3.1}$$

are rational numbers (we have replaced k by $k-1$ in (2.6)). It would be nice to find a continuous p-adic function $\zeta_p: Z_p \longrightarrow Q_p$

which agrees with ζ on all $1-k$. Since the set $\{1-k\}$ is dense in Z_p, there can be at most one such ζ_p. Such a ζ_p exists if and only if

$$k_1 \text{ close to } k_2 \text{ p-adically} \Longrightarrow \zeta(1-k_1) \text{ close to } \zeta(1-k_2) \text{ p-adically.}$$

This is not the case, however, and we must first modify the zeta function.

We define a new complex analytic function by setting

$$\zeta^*(s) = \sum_{p \nmid n} n^{-s} = (1 - p^{-s})\,\zeta(s)$$

for $\mathrm{Re}\ s > 1$ (and for other $s \in C$ by analytic continuation). ζ^* is obtained from ζ in a similar way to how the Artin-Hasse exponential was obtained from e^x in §I.3 (see the identity (3.2) in Chapter I; the terms with $p|n$ are omitted to define $E_p(x)$). This procedure is often called "removing the Euler factor at p", because

$$\zeta^*(s) = (1-p^{-s})\zeta(s) = (1-p^{-s})\prod_{\ell} \frac{1}{1-\ell^{-s}} = \prod_{\ell \neq p} \frac{1}{1-\ell^{-s}}.$$

There is yet another way to view ζ^*. Let us return to the measures μ_ε. It is easy to see that there does not exist a translation-invariant (bounded) p-adic measure, i.e., a μ on Z_p such that

$$\mu(a_1 + p^N Z_p) = \mu(a_2 + p^N Z_p) \quad \text{for all } a_1, a_2.$$

However, the measures μ_ε on Z_p (for any $\varepsilon \notin D_1(1^-)$) have the closest possible property, namely:

$$\mu_{\varepsilon^p}(a + p^N Z_p) = \mu_\varepsilon(ap + p^{N+1} Z_p),$$

as follows trivially from the definition. This implies that for any continuous function f on Z_p

$$\int_{pZ_p} f(x)\, d\mu_\varepsilon(x) = \int_{Z_p} f(px)\, d\mu_{\varepsilon^p}(x) \tag{3.2}$$

(where for $U \subset X$, $\int_U f$ of course means $\int (f|_U$ extended by zero to $X - U)$). Now let $\varepsilon^r = 1$, $p \nmid r$. Since raising to the p-th power permutes r-th roots of one, we have (where we again let $\mu =$

$\Sigma\, \mu_\varepsilon$):

$$\int_{Z_p^*} x^{k-1}\, d\mu(x) = \left(\int_{Z_p} - \int_{pZ_p}\right) x^{k-1}\, d\mu(x)$$

$$= \int_{Z_p} x^{k-1}\, d\mu(x) - \int_{Z_p} (px)^{k-1}\, d\mu(x)$$

$$= (1 - p^{k-1}) \int_{Z_p} x^{k-1}\, d\mu(x).$$

Dividing by r^k-1, we obtain by (3.1)

$$\frac{1}{r^k - 1} \int_{Z_p^*} x^{k-1}\, d\mu(x) = (1 - p^{k-1})\, \zeta(1-k) = \zeta^*(1-k). \qquad (3.3)$$

Thus, removing the Euler factor is equivalent to restricting the domain of integration from Z_p to Z_p^*.

Now suppose that two values k_1 and k_2 are close p-adically, and are also in the same congruence class mod p-1, that is, suppose that $k_1 - k_2 = (p-1)p^N m$, $m \in Z$. Then we compare the integrand in (3.3) for k_1 and k_2:

$$\frac{x^{k_1-1}}{x^{k_2-1}} = x^{k_1-k_2} = \left(x^{p-1}\right)^{p^N m}.$$

But for $x \in Z_p^*$, $x^{p-1} \equiv 1 \pmod p$ (because $a^{p-1} = 1$ for $a \in F_p^*$), and it is easy to see (using the binomial expansion) that

$$\left(x^{p-1}\right)^{p^N m} \equiv 1 \pmod{p^{N+1}}. \qquad (3.4)$$

Thus, x^{k_1-1} and x^{k_2-1} are close together p-adically. Hence, their integrals over the compact set Z_p^* are also close together; in fact, it is easy to see that

$$\int_{Z_p^*} x^{k_1-1}\, d\mu(x) \equiv \int_{Z_p^*} x^{k_2-1}\, d\mu(x) \pmod{p^{N+1}}.$$

If we further assume that $k_1 \not\equiv 0 \pmod{p-1}$, and if we take r to be a primitive (p-1)-th root of one modulo p (so that $p \nmid r^{k_1}-1$), then we have: $1/(r^{k_1} - 1) \equiv 1/(r^{k_2} - 1) \pmod{p^{N+1}}$. Multiplying these two congruences and using (3.3) and (3.1), we obtain the

Kummer congruences. *If* $k_1 \equiv k_2 \pmod{(p-1)p^N}$ *and* $(p-1)\nmid k_1$, *then*

$$(1 - p^{k_1-1}) \frac{B_{k_1}}{k_1} \equiv (1 - p^{k_2-1}) \frac{B_{k_2}}{k_2} \pmod{p^{N+1}},$$

where both sides of the congruence are rational numbers in Z_p *(i.e., without* p *in the denominator).*

Thus, the Kummer congruences, which were originally thought to be merely a number theoretic curiosity, are now seen to arise naturally from the simple fact that: if two functions are close together, then their integrals over a compact set are also close together.

We can now define the p-adic zeta function $\zeta_p(s)$ by letting $1-k$ approach s p-adically, but *fixing* a class modulo $p-1$, i.e., fixing $k_0 \in \{0, 1, \ldots, p-2\}$ and only choosing k which are congruent to $k_0 \pmod{p-1}$. Thus, we define

$$\begin{aligned}
\zeta_{p,k_0}(s) &= \lim_{1-k \to s,\ k \equiv k_0 \pmod{p-1}} \zeta^*(1-k) \\
&= \lim_{1-k \to s,\ k \equiv k_0 \pmod{p-1}} \frac{1}{r^k - 1} \int_{Z_p^*} x^{k-1}\, d\mu(x) \\
&= \frac{1}{\langle r\rangle^{1-s}\, \omega(r)^{k_0} - 1} \int_{Z_p^*} \langle x\rangle^{-s}\, \omega(x)^{k_0-1}\, d\mu(x),
\end{aligned}$$

where ω is the locally constant function on Z_p^* which takes a p-adic integer to the Teichmüller representative of its first digit, and as before $\langle x\rangle = x/\omega(x) \equiv 1 \pmod{p}$. (Thus, $\langle x\rangle^{-s}$ is well defined for p-adic s, see (3.4).) ζ_p is a p-adic function with $p-1$ "branches" ζ_{p,k_0} for $k_0 = 0, 1, \ldots, p-2$.

Remark. The classical Mellin transform of a measure $\mu = f(x)dx$ is the function

$$g(s) = \int_0^\infty x^s\, f(x)dx.$$

For example, the gamma function is defined as the Mellin transform of $e^{-x}dx/x$:

$$\Gamma(s) = \int_0^\infty e^{-x}\, x^{s-1}\, dx.$$

Thus, ζ_p is the p-adic "Mellin-Mazur transform" of the measure $\mu = \sum_{\varepsilon^r=1,\ \varepsilon\neq 1} \mu_\varepsilon = \mu_{\text{Mazur},1/r}$. Strangely, the p-adic Γ-function, which we shall soon study, is not any type of p-adic Mellin transform, so far as we know.

4. p-adic Dirichlet L-functions

A Dirichlet character $\chi: (Z/dZ)^* \longrightarrow C^*$ takes values in a finite (cyclotomic) extension K of Q. Recall that we can consider K to be imbedded in Ω_p if we choose a prime ideal P of K dividing p and take the completion of K in the P-adic topology: $\iota_p: K \hookrightarrow \Omega_p$. We shall still use the letter χ for $\iota_p \circ \chi$, so χ denotes either a complex or p-adic valued character.

A Dirichlet character χ is said to be primitive of conductor d if there is no character $\chi': (Z/d'Z)^* \longrightarrow C^*$, d' a proper divisor of d, such that $\chi(n) = \chi'(n)$ for all n prime to d; equivalently, χ is primitive if it is not constant on any subgroup $\{x \mid x \equiv 1 \pmod{d'}\}$ in $(Z/dZ)^*$.

If χ_1 and χ_2 are two primitive Dirichlet characters of conductor d_1 and d_2, respectively, then $\chi_1\chi_2$ denotes the primitive Dirichlet character such that $\chi_1\chi_2(n) = \chi_1(n)\chi_2(n)$ whenever n and d_1d_2 have no common factor. This is not the same as the character $n \mapsto \chi_1(n)\chi_2(n)$, which is often imprimitive. For example, if $\chi_2 = \overline{\chi_1}$ is the conjugate character, then $\chi_1\chi_2$ is identically 1, while $\chi_1(n)\chi_2(n) = 0$ if $\text{g.c.d.}(n,d_1) > 0$. Note that the conductor of $\chi_1\chi_2$ divides the least common multiple of d_1, d_2.

If χ is a primitive Dirichlet character of conductor d with values in Ω_p, we let $\chi_k = \chi\omega^{-k}$, where $\omega: n \mapsto \omega(n)$ is the Teichmüller character, which has conductor p. Clearly, the conductor of χ_k is pd if $p \nmid d$, and is either d or d/p if $p \mid d$.

Let χ be a primitive Dirichlet character of conductor d. We use the relation (2.5) for $f = \chi$:

$$L(1-k,\chi,z) = \int_X x^{k-1} \chi(x)\, d\mu_z(x).$$

(We have replaced k by $k-1$ in (2.5).) We want p-adically to interpolate this function of k, i.e., to let k approach $1-s \in Z_p$ and get an Ω_p-valued function $L_p(s,\chi,z)$. To do this, we must first make two modifications: (1) "remove the Euler factor" by restricting the integral to

$$X^* \underset{\text{def}}{=} \bigcup_{0<a<dp,\ p\nmid a} a + dpZ_p$$

(X^* is the inverse image of Z_p^* under the "forget mod d information" map); (2) replace x by $\langle x\rangle = x/\omega(x)$ in x^k. We thus define

$$\begin{aligned} L_p(1-k,\chi,z) &\underset{\text{def}}{=} \int_{X^*} \langle x\rangle^k/x\ \chi(x)\, d\mu_z(x) \\ &= \int_{X^*} \langle x\rangle^{k-1} \chi_1(x)\, d\mu_z(x) \\ &= \int_{X^*} x^{k-1} \chi_k(x)\, d\mu_z(x) \\ &= \int_X x^{k-1}\chi_k(x)d\mu_z(x) - \int_X (px)^{k-1} \chi_k(px)d\mu_{z^p}(x) \end{aligned}$$

(see (3.2); the argument is the same for X as for Z_p). Bringing the p outside the second integral and using the above expression for $L(1-k,\chi,z)$, we conclude that

$$L_p(1-k,\chi,z) = L(1-k,\chi_k,z) - p^{k-1}\chi_k(p)L(1-k,\chi_k,z^p). \tag{4.1}$$

We thus have the following

Proposition. *For χ a character of conductor d, the continuous function from Z_p to Ω_p*

$$L_p(s,\chi,z) \underset{\text{def}}{=} \int_{X^*} \langle x\rangle^{-s} \chi_1(x)\, d\mu_z(x)$$

interpolates the values $L(1-k,\chi_k,z) - p^{k-1}\chi_k(p)L(1-k,\chi_k,z^p)$.

This proposition can be used to prove the following theorem.

Theorem (Kubota-Leopoldt [58] and Iwasawa [41]). *There exists a unique p-adic continuous (except for a pole at 1 when χ is* the trivial character) function $L_p(s,\chi)$, $s \in Z_p$, such that

$$L_p(1-k,\chi) = (1 - \chi_k(p)p^{k-1})L(1-k,\chi_k). \tag{4.2}$$

Proof. Let $r > 1$ be an integer prime to pd, and let $z^r = 1$, $z \neq 1$. We first note that the ordinary classical L-function can be recovered from the "twisted" L-function $L(s,\chi,z)$ by means of the following relation, which follows immediately from the definitions:

$$\sum_{z^r=1,\ z\neq 1} L(s,\chi,z) = (r^{1-s}\chi(r) - 1)\ L(s,\chi), \quad s \in C. \tag{4.3}$$

Using this for $s = 1-k$ and summing (4.1) over nontrivial r-th roots of unity z (which are only permuted by $z \mapsto z^p$), we have

$$\begin{aligned} \sum_{z^r=1,\ z\neq 1} L_p(1-k,\chi,z) &= (1-\chi_k(p)p^{k-1}) \sum_{z^r=1,\ z\neq 1} L(1-k,\chi_k,z) \\ &= (r^k\chi_k(r)-1)(1-\chi_k(p)p^{k-1})\ L(1-k,\chi_k) \\ &= (\langle r\rangle^k\chi(r)-1)(1-\chi_k(p)p^{k-1})L(1-k,\chi_k) \end{aligned} \tag{4.4}$$

So we define

$$\begin{aligned} L_p(s,\chi) &\underset{\text{def}}{=} \frac{1}{\langle r\rangle^{1-s}\chi(r) - 1} \sum_{z^r=1,\ z\neq 1} L_p(s,\chi,z) \\ &= \frac{1}{\langle r\rangle^{1-s}\chi(r) - 1} \int_{X^*} \langle x\rangle^{-s}\ \chi_1(x)\ d\mu(x), \end{aligned} \tag{4.5}$$

where μ is the sum of μ_z over all z with $z^r = 1$, $z \neq 1$. The equality (4.2) in the theorem now follows from (4.4). The continuity of $L_p(s,\chi)$ (more precisely, local analyticity) follows because we are taking the integral of a continuous (actually, analytic) function of s and then dividing by an expression which can only vanish if $s = 1$ and $\chi(r) = 1$; r can be chosen so that $\chi(r) \neq 1$ unless χ is trivial. This concludes the proof of the theorem.

Notice that the function $\zeta_{p,k_0}(s)$ we defined in §3 is precisely $L_p(s,\omega^{k_0})$. Also note that, while it was necessary to choose r in order to construct both $\zeta_{p,k_0}(s)$ and $L_p(s,\chi)$, these functions are in fact independent of r.

5. Leopoldt's formula for $L_p(1,\chi)$.

Recall [13] the classical formula for $L(1,\chi)$, which can be derived by Fourier inversion on the group $G = Z/dZ$. Let ζ be a fixed primitive d-th root of 1, and define

$$\hat{f}(a) = \sum_{b\in G} f(b)\,\zeta^{-ab}$$

for a function f on G. Then

$$f(b) = \frac{1}{d}\sum_{a\in G} \hat{f}(a)\,\zeta^{ab}. \tag{5.1}$$

Applying Fourier inversion (5.1) to $f_s(b) = \Sigma_{n\equiv b \ (\mathrm{mod}\ d)}\, n^{-s}$ (suppose Re $s > 1$) and using the definition of $L(s,\chi)$ and $L(s,1,z) = \Sigma\, z^n n^{-s}$, we have

$$\begin{aligned} L(s,\chi) &= \sum_{0\le b<d} \chi(b)\, f_s(b) = \frac{1}{d}\sum_{a,b} \chi(b)\,\hat{f}_s(a)\,\zeta^{ab} \\ &= \frac{1}{d}\sum_j \chi(j)\zeta^j \sum_a \bar{\chi}(a)\,\hat{f}_s(a) \qquad (\text{where } j = ab) \\ &= \frac{g_\chi}{d}\sum_a \bar{\chi}(a)\, L(s,1,\zeta^{-a}), \end{aligned}$$

where $g_\chi = \Sigma\chi(j)\zeta^j$ is the Gauss sum. Letting $s\to 1$ and noting that $L(1,1,z) = -\log(1-z)$, we obtain

$$L(1,\chi) = -\frac{g_\chi}{d}\sum_{0<a<d} \bar{\chi}(a)\log(1-\zeta^{-a}). \tag{5.2}$$

We now proceed to the p-adic case.

Theorem (Leopoldt [64]).

$$L_p(1,\chi) = -\left(1-\frac{\chi(p)}{p}\right)\frac{g_\chi}{d}\sum_{0<a<d} \bar{\chi}(a)\, \ln_p(1-\zeta^{-a}). \tag{5.3}$$

The purpose of this section is to prove this theorem.

Note that (5.3) differs from (5.2) in two respects: the expected "removal of the Euler factor", giving the term $(1-\chi(p)/p)$; and the replacement of log by $\ln_p$. The validity of the p-adic formula (5.3) might seem surprising at first because of the replacement of $\log(1-\zeta^{-a})$ by $\ln_p(1-\zeta^{-a})$, since the formal series for the former does not converge p-adically, i.e., we need properties (2) and (3) of the proposition in §I.3 in order to evaluate

$\ln_p(1-\zeta^{-a})$. But the proof of Lemma 1 below shows that the same formal series, along with the p-adic version of analytic continuation, really do lie behind Leopoldt's formula, despite first impressions.

Lemma 1. If $z \notin D_1(1^-)$ and μ_z is the measure on $\mathbb{Z}_p$ given by $\mu_z(a + p^N\mathbb{Z}_p) = z^a/(1-z^{p^N})$, then

$$\int_{\mathbb{Z}_p^*} \frac{1}{x}\, d\mu_z(x) = -\frac{1}{p}\ln_p\frac{(1-z)^p}{1-z^p}.$$

Proof. If $|z|_p < 1$, then the left side is (the $'$ denotes omission of indices divisible by p):

$$\lim_{N\to\infty} \sum_{0<j<p^N}{}' \frac{z^j}{j}\,\frac{1}{1-z^{p^N}} = \lim_{N\to\infty}\left(\sum_{0<j<p^N}\frac{z^j}{j} - \sum_{0<j<p^{N-1}}\frac{z^{pj}}{pj}\right)$$

$$= -\frac{1}{p}\left(\ln_p(1-z)^p - \ln_p(1-z^p)\right).$$

We now use analytic continuation to extend the equality from $|z|_p < 1$ to all $z \notin D_1(1^-)$. As we remarked at the end of §I.3, a function is said to be Krasner analytic on the complement of $D_1(1^-)$ if it is a uniform limit of rational functions with poles in $D_1(1^-)$. The basic fact we need about such functions (see [57]) is that if two Krasner analytic functions on the complement of $D_1(1^-)$ are equal on a disc, then they are equal everywhere on the complement of $D_1(1^-)$. Thus, if we show that the two sides of Lemma 1 are each Krasner analytic functions of z on the complement of $D_1(1^-)$, then, since they are equal on the disc $|z|_p < 1$, they must be equal for all $z \notin D_1(1^-)$.

Note that by writing

$$\frac{(1-z)^p}{1-z^p} = \begin{cases} 1 + \dfrac{1}{1-z^p}\displaystyle\sum_{0<j<p}\binom{p}{j}(-z)^j, & \text{if } |z|_p \le 1,\ z \notin D_1(1^-); \\[2ex] 1 + \dfrac{1}{1-z^{-p}}\displaystyle\sum_{0<j<p}\binom{p}{j}(-z)^{-j}, & \text{if } |z|_p > 1, \end{cases}$$

we see that $z \notin D_1(1^-) \implies (1-z)^p/(1-z^p) \in D_1(1^-)$ (in fact, its distance from 1 is $\leq 1/p$). Hence, the right side of the equality in the lemma is the uniform limit of the rational functions (with poles in $D_1(1^-)$)

$$\frac{1}{p}\sum_{j=1}^{N}\frac{(-1)^j}{j}\left(\frac{(1-z)^p}{1-z^p}-1\right)^j,$$

and the left side is the uniform limit of the rational functions (with poles in $D_1(1^-)$)

$$\sum_{0<j<p^N}{}' \frac{z^j}{j}\,\frac{1}{1-z^{p^N}}.$$

This concludes the proof of the lemma.

Lemma 1 will be applied when z is a root (but not a p^N-th root) of unity.

Remarks. 1. If z is a $(p-1)$-th root of 1, the right side of Lemma 1 becomes $-(1-1/p)\ln_p(1-z)$. For example, setting $z = -1$ gives the following p-adic limit for $\ln_p 2$:

$$\ln_p 2 = -\frac{p}{2(p-1)}\lim_{N\to\infty}\sum_{0<j<p^N}{}' \frac{(-1)^j}{j}.$$

2. Lemma 1 is the key step in our proof of Leopoldt's formula for $L_p(1,\chi)$. As mentioned before, the subtlety in Leopoldt's formula is that $\ln_p(1-z)$ is not given by the same formal series as $\log(1-z)$, since z is outside the disc of convergence of $\ln_p(1-z)$. However, Lemma 1 shows that if we "correct by the Frobenius" in the Dwork style (see, e.g., [28]), i.e., if we replace $(1-z)$ by $(1-z)^p/(1-z^p)$, then the resulting series _is_ globally analytic out to roots of unity. (We shall see further examples of "correcting by the Frobenius", e.g., in §6.) The effect of this step on the formula for $L_p(1,\chi)$ is to bring out the Euler factor $(1 - \chi(p)/p)$, as we shall see below (in (5.4)).

The other ingredient in the proof of Leopoldt's formula is the analog of the Fourier inversion (5.1) used in the classical case.

Lemma 2. *Suppose that* χ *is a primitive Dirichlet character mod* d, ζ *is a fixed primitive d-th root of unity,* $g_\chi = \Sigma\chi(j)\zeta^j$, $z \neq 1$ *is an r-th root of unity, where* r *is prime to* pd, *and* $f: X \longrightarrow \Omega_p$ *is any continuous function. Then*

$$\int_X \chi\, f\, d\mu_z = \frac{g_\chi}{d} \sum_{0\le a<d} \bar{\chi}(a) \int_X f\, d\mu_{\zeta^{-a}z}.$$

To prove Lemma 2, by linearity and continuity of both sides it suffices to prove it when f is the characteristic function of $j + dp^N Z_p$, i.e., to prove that $\chi(j)z^j/(1-z^{dp^N}) = \frac{g_\chi}{d} \Sigma\, \bar{\chi}(a) z^j \zeta^{-aj}/(1-z^{dp^N})$. But this reduces to: $g_\chi \bar{g}_\chi = d$.

Proof of the theorem. We first prove an analogous formula for the "twisted" $L_p(1,\chi,z)$.

Note that, if $f: X \longrightarrow \Omega_p$ comes from pulling back a function (also denoted f) on Z_p using the projection ("forget mod d information") from X to Z_p, then we can replace X by Z_p in $\int_X f\, d\mu_z$, where μ_z on Z_p is defined by the same formula as on X with d replaced by 1: $\mu_z(j + p^N Z_p) = z^j/(1-z^{p^N})$. To see this, one reduces to the case when f is the pull-back of the characteristic function of $j + p^N Z_p$, which is checked easily.

Applying Lemma 2 and the preceding remark to the function $f(x) = \frac{1}{x}\cdot$(characteristic function of X^*), we obtain

$$L_p(1,\chi,z) = \int_{X^*} \frac{\chi(x)}{x} d\mu_z = \frac{g_\chi}{d} \sum_{0\le a<d} \bar{\chi}(a) \int_{Z_p^*} \frac{1}{x}\, d\mu_{\zeta^{-a}z}$$

$$= -\frac{g_\chi}{d} \sum_{0\le a<d} \bar{\chi}(a)\, \frac{1}{p} \ln_p \frac{(1-\zeta^{-a}z)^p}{1-(\zeta^{-a}z)^p} \quad \text{by Lemma 1}$$

$$= -\frac{g_\chi}{d}\left(1 - \frac{\chi(p)}{p}\right) \sum_{0<a<d} \bar{\chi}(a) \ln_p(1 - \zeta^{-a}z). \tag{5.4}$$

Since $r > 1$ is any integer prime to pd, we may choose r so that $\chi(r) \neq 1$ and then use (4.5) with $s = 1$ to express $L_p(1,\chi)$ in terms of the $L_p(1,\chi,z)$. We obtain

$$L_p(1,\chi) = \frac{1}{\chi(r)-1} \sum_{z^r=1,\ z\neq 1} L_p(1,\chi,z)$$

$$= -\frac{g_\chi}{d}\left(1-\frac{\chi(p)}{p}\right)\left[\frac{1}{\chi(r)-1}\sum_{0<a<d}\overline{\chi}(a)\sum_{z^r=1,\ z\neq 1}\ln_p(1-\zeta^{-a}z)\right].$$

Since the inner summation is equal to $\ln_p(1-\zeta^{-ar}) - \ln_p(1-\zeta^{-a})$, the term in square brackets is immediately seen to equal $\Sigma_{0<a<d}\ \overline{\chi}(a)\ln_p(1-\zeta^{-a})$, as desired.

This completes the proof of Leopoldt's formula.

We shall later prove one more formula for the p-adic L-function, which relates its behavior at 0 to the p-adic gamma function. But first we take up the p-adic gamma and log gamma functions.

6. The p-adic gamma function

First recall some properties of the classical gamma function:

(1) It is a meromorphic function on the complex plane with poles at $0, -1, -2, -3, \ldots$.

(2) $\Gamma(x+1) = x\Gamma(x)$, $\Gamma(k) = (k-1)!$

(3) $$\Gamma(x)\cdot\Gamma(1-x) = \frac{\pi}{\sin(\pi x)} = \frac{2\pi i}{e^{2\pi i x}-1}\, e^{\pi i x}$$

(4) $$\Gamma(x) = \lim_{n\to\infty} \frac{1\cdot 2\cdots(n-1)}{x(x+1)\cdots(x+n-1)}\, n^x$$

(5) Gauss multiplication formula: for $m = 1, 2, \ldots$

$$\prod_{h=0}^{m-1} \Gamma\left(\frac{x+h}{m}\right) = (2\pi)^{(m-1)/2}\, m^{\frac{1}{2}-x}\, \Gamma(x),$$

for example $(x = 1)$

$$\prod_{h=1}^{m-1} \Gamma\left(\frac{h}{m}\right) = (2\pi)^{(m-1)/2}\, m^{-1/2},$$

so that if we divide these two equations we obtain

$$\frac{\prod_{h=0}^{m-1} \Gamma\left(\frac{x+h}{m}\right)}{\Gamma(x)\prod_{h=1}^{m-1}\Gamma\left(\frac{h}{m}\right)} = m^{1-x}. \tag{6.1}$$

We now proceed to the p-adic theory. For simplicity, we shall assume that $p \neq 2$. (Minor modifications are sometimes needed when $p = 2$.)

Proceeding naively, we would like to construct a function $\Gamma_p(s)$ on Z_p which interpolates $\Gamma(k) = \Pi_{j<k}\, j$, i.e., so that $\Gamma(k)$ approaches $\Gamma_p(s)$ as k runs through any sequence of positive integers which approaches s p-adically. However, $\Gamma(k)$ is divisible by a large power of p for large k; hence $\Gamma_p(s)$ would have to be identically zero, which is useless. So, as in the case of the zeta function, we must modify the values $\Gamma(k)$.

We might improve the situation if we eliminate the j's which are divisible by p, much as we "removed the Euler factor at p" from $\zeta(s)$ to get $\zeta^*(s)$. Thus, suppose we take $\Gamma^*(k) = \Pi_{j<k,\ p\nmid j}\, j$, which is an integer prime to p. Now we can find a continuous function Γ_p on Z_p which agrees with Γ^* on positive integers if and only if

k_1 close to k_2 p-adically $\Longrightarrow$ $\Gamma^*(k_1)$ close to $\Gamma^*(k_2)$ p-adically,

i.e., if and only if $\Gamma^*(k_2)/\Gamma^*(k_1) \equiv 1 \pmod{p^N}$ with N large whenever $k_2 - k_1$ is highly divisible by p. To check this, take for example $k_2 = k_1 + p^n$. But it is easy to show that in that case the quotient $\Gamma^*(k_2)/\Gamma^*(k_1) = \Pi_{k_1 \le j < k_2,\ p\nmid j}\, j$ is congruent to $-1 \bmod p^n$. (Namely, in any finite abelian group G we have $\Pi_{g\in G}\, g = \Pi_{g=g^{-1}}\, g$; apply this to $G = (Z/p^nZ)^*$.) So the sign is wrong, and we have to make one final modification. We define

$$\Gamma_p(k) = (-1)^k \prod_{j<k,\ p\nmid j} j, \qquad \Gamma_p(s) = \lim_{k\to s} \Gamma_p(k). \tag{6.2}$$

Using the generalized Wilson's theorem cited above:

$$\prod_{k\le j<k+p^n,\ p\nmid j} j \equiv -1 \pmod{p^n},$$

it is simple to check that the limit in (6.2) exists, is independent of how k approaches s, and determines a continuous function on Z_p with values in Z_p^*.

We verify some basic properties of Γ_p:

(1) $\Gamma_p(0) = 1$, and $\dfrac{\Gamma_p(x+1)}{\Gamma_p(x)} = \begin{cases} -x & \text{if } x \in Z_p^*; \\ -1 & \text{if } x \in pZ_p. \end{cases}$

To prove the second equality, since both sides are continuous on Z_p, it suffices to prove equality for the positive integers $x = k$ (which are dense in Z_p), and then it is obvious from (6.2). Using (6.2) and this equality, we can compute the first few values $\Gamma_p(2) = 1$, $\Gamma_p(1) = -1$, $\Gamma_p(0) = 1$.

(2) For $x \in Z_p$, write $x = x_0 + px_1$, where $x_0 \in \{1,\ldots,p\}$ is the first digit in x, unless $x \in pZ_p$, in which case $x_0 = p$. Then we have

$$\Gamma_p(x)\cdot\Gamma_p(1-x) = (-1)^{x_0}.$$

In fact, to show that the continuous function $f(x) = (-1)^{x_0}\Gamma_p(x)\cdot\Gamma_p(1-x)$ equals 1 on Z_p, it suffices to show that $f(k) = 1$ for positive integers k. Clearly $f(1) = 1$, and a simple verification using property (1) shows that $f(k+1)/f(k) = 1$.

(3) For any positive integer m, $p \nmid m$, we have (here x_0 and x_1 are as in property (2)):

$$\frac{\prod_{h=0}^{m-1} \Gamma_p\left(\frac{x+h}{m}\right)}{\Gamma_p(x) \prod_{h=1}^{m-1} \Gamma_p\left(\frac{h}{m}\right)} = m^{1-x_0}\left(m^{-(p-1)}\right)^{x_1}. \tag{6.3}$$

(Note: Since $m^{-(p-1)} \equiv 1 \pmod p$, it follows that $\left(m^{-(p-1)}\right)^{x_1}$ is a well-defined function of p-adic x_1.) To prove property (3), let $f(x)$ be the left side and $g(x)$ the right side of (6.3). f and g are continuous, and $f(1) = 1 = g(1)$. Next,

$$\frac{f(x+1)}{f(x)} = \frac{\Gamma_p(x)}{\Gamma_p(x+1)} \cdot \frac{\Gamma_p\left(\frac{x}{m}+1\right)}{\Gamma_p(x/m)} = \begin{cases} 1/m & \text{if } x \in Z_p^*; \\ 1 & \text{if } x \in pZ_p, \end{cases}$$

while

$$\frac{g(x+1)}{g(x)} = \begin{cases} 1/m & \text{if } x \in Z_p^*, \text{ since then } (x+1)_0 = x_0+1, (x+1)_1 = x_1; \\ 1 & \text{if } x \in pZ_p, \text{ since then } (x+1)_0 = x_0-(p-1), \\ & \qquad (x+1)_1 = x_1+1. \end{cases}$$

This proves property (3).

We now discuss an interesting special case of property (3). Suppose that $x = \frac{r}{p-1}$ is a rational number between 0 and 1 whose denominator divides $p-1$. Then $x_0 = p-r$, $x_1 = \frac{r}{p-1} - 1 = (1-x_0)/(p-1)$. Note that the left side $f(x)$ of (6.3) is congruent to $m^{1-x_0} = m^{(1-x)(1-p)}$ mod p. In addition,

$$f(x)^{p-1} = \left(m^{p-1}\right)^{(1-x_0 - x_1(p-1))} = 1.$$

Thus, $f(x)$ is the $(p-1)$-th root of 1 congruent to $m^{(1-x)(1-p)}$ mod p:

$$f(x) = \omega\left(m^{(1-x)(1-p)}\right).$$

Now the classical expression $f_{cl}(x)$ which is obtained from $f(x)$ by replacing Γ_p by Γ, is equal to $m^{1-x} \in Q(\sqrt[p-1]{m})$. Let $K = Q(\xi)$, where ξ is a fixed primitive $(p-1)$-th root of unity. Then $\mathrm{Gal}(K(\sqrt[p-1]{m})/K) \cong Z/(p-1)Z$, with $\sigma_a: \sqrt[p-1]{m} \mapsto \xi^a \sqrt[p-1]{m}$ for $a \in Z/(p-1)Z$. Choose a prime ideal P of $K(\sqrt[p-1]{m})$ which divides p. Then P determines an imbedding $\iota_P: K(\sqrt[p-1]{m}) \hookrightarrow Q_p(\sqrt[p-1]{m})$. There exists a unique "Frobenius element" Frob $\in \mathrm{Gal}(K(\sqrt[p-1]{m})/K)$ such that $\mathrm{Frob}(x) \equiv x^p \pmod{P}$ for every algebraic integer x in $K(\sqrt[p-1]{m})$.

We then have for $x = r/(p-1)$

$$f_{cl}(x)^{1-\mathrm{Frob}} = \frac{m^{1-x}}{\mathrm{Frob}\, m^{1-x}} \equiv m^{(1-x)(1-p)} \pmod{P};$$

and, since elements of $\mathrm{Gal}(K(\sqrt[p-1]{m})/K)$ multiply m^{1-x} by roots of unity, we conclude that

$$\iota_p\left(f_{cl}(x)^{1-\mathrm{Frob}}\right) = \omega\left(m^{(1-x)(1-p)}\right) = f_{p\text{-adic}}(x).$$

This phenomenon -- that a classical expression raised to the 1-Frob, where Frob is a p-th power type map, can be identified with a p-adic analog of the classical expression -- occurs in other contexts. For example, let

$$E:\ y^2 = x(x-1)(x-\lambda),\quad \lambda \in \mathbf{Z},$$

be an elliptic curve whose reduction

$$\overline{E}:\ y^2 = x(x-1)(x-\overline{\lambda}),\quad \overline{\lambda} \in \mathbf{Z}/p\mathbf{Z},$$

is nonsingular. Further suppose that $\overline{E}$ is not "supersingular" (which will be the case if $\overline{\lambda}$ is not a root of the polynomial $\sum_{n=0}^{(p-1)/2} \binom{(p-1)/2}{n}^2 \overline{\lambda}^n$). It is known [67,43] that the period f of the holomorphic differential dx/y on the Riemann surface (torus) E, as a function of the parameter λ, satisfies the differential equation

$$\lambda(1-\lambda)f'' + (1-2\lambda)f' - \tfrac{1}{4}f = 0,$$

whose solution bounded at zero is the hypergeometric series

$$f(\lambda) = \sum_{n=0}^{\infty} \binom{-1/2}{n}^2 \lambda^n \in \mathbf{Q}[[\lambda]].$$

Although when we consider $f(\lambda)$ as a p-adic series it converges only on $D(1^-)$, it turns out that the power series $\Theta = f^{1-\mathrm{Frob}}$ defined by $\Theta(\lambda) = f(\lambda)/f(\lambda^p)$ converges on $D(\gamma)$ for some $\gamma > 1$. Now it is well known that the zeta-function of $\overline{E}$

$$Z(\overline{E}/\mathbf{F}_p) = \exp\sum\left(\frac{(\text{number of } \mathbf{F}_{p^n}\text{-points on } \overline{E})}{n}\right)T^n$$

is of the form

$$Z(\overline{E}/\mathbf{F}_p) = \frac{(1-\alpha T)(1-pT/\alpha)}{(1-T)(1-pT)},$$

where $\mathrm{ord}_p\alpha = 0$. Dwork [43] proved the following formula for α:

$$\alpha = \Theta(\omega(\overline{\lambda})).$$

Thus, α can be thought of as a sort of "p-adic period".

In Chapter III we shall study another analogy between classical formulas for periods of a curve considered over $\mathbf{C}$ and p-adic formulas for the roots of the zeta function of the curve considered over a finite field.

This concludes our discussion of elementary (easily proved) properties of Γ_p. In Chapter III we shall prove an algebraicity result for certain Γ_p values and products of values, for example, the algebraicity of $\Gamma_p\left(\frac{r}{d}\right)$ if $d|p-1$. (More precisely: $\left(\Gamma_p\left(\frac{r}{d}\right)\right)^d \in Q(\sqrt[d]{1})$.) But the proof of this fact uses p-adic cohomology. It would be interesting to find an elementary proof of this algebraicity. After all, the assertion can be stated very simply (here we write $\frac{r}{d} = 1 - \frac{s}{p-1}$):

$$\lim_{n\to\infty} \frac{(s+sp+\ldots+sp^n+sp^{n+1})!}{(s+sp+\ldots+sp^n)!\; p^{s+sp+\ldots+sp^n}} \in Z_p^* \text{ is algebraic over } Q.$$

For example, the theorem in §III.6 will give us the following 5-adic formula:

$$\Gamma_5\left(\frac{1}{4}\right)^4 = \lim_{n\to\infty}\left(\frac{(3+3\cdot 5+\ldots+3\cdot 5^n+3\cdot 5^{n+1})!}{(3+3\cdot 5+\ldots+3\cdot 5^n)!\; 5^{3+3\cdot 5+\ldots+3\cdot 5^n}}\right)^4$$

$$= 3 + 4\omega(2) = 3 + 4\sqrt{-1} \in Z_5^*$$

(where $\sqrt{-1} = \omega(2) = 2 + 1\cdot 5 + 2\cdot 5^2 + 1\cdot 5^3 + 3\cdot 5^4 + \ldots \in Z_5$) and the following 7-adic formula:

$$-\Gamma_7\left(\frac{1}{3}\right)^3 = \lim_{n\to\infty}\left(\frac{(4+\ldots+4\cdot 7^n+4\cdot 7^{n+1})!}{(4+\ldots+4\cdot 7^n)!\; 7^{4+\ldots+4\cdot 7^n}}\right)^3$$

$$= \frac{-1+3\sqrt{-3}}{2} = 1 + 3\omega(4) \in Z_7^*.$$

No elementary proof is known for either of these equalities.

7. The p-adic log gamma function

We start by describing another approach to the p-adic zeta and L-functions, which was the original point of view of Kubota and Leopoldt [58].

It is not hard to prove the following p-adic formula for the k-th Bernoulli number (see [41]):

$$B_k = \lim_{n\to\infty} p^{-n} \sum_{0\le j<p^n} j^k. \tag{7.1}$$

More generally, if $f: Z \to \Omega_p$ has period d, and if $B_{k,f}$ is defined for p-adic valued f in the same way as for complex valued f (see (1.1)), then we have

$$B_{k,f} = \lim_{n\to\infty} \frac{1}{dp^n} \sum_{0\le j<dp^n} f(j)\, j^k. \tag{7.2}$$

The simplest examples of (7.1) are:

$$\lim p^{-n} \sum_{j<p^n} j = \lim p^{-n} \frac{p^n(p^n-1)}{2} = -\frac{1}{2} = B_1;$$

$$\lim p^{-n} \sum_{j<p^n} j^2 = \lim p^{-n} \frac{p^n(p^n-1)(2p^n-1)}{6} = \frac{1}{6} = B_2.$$

This type of limit $\lim\limits_{n\to\infty} p^{-n} \sum\limits_{0\le j<p^n} f(j)$ can be used for other $f(x)$ besides $f(x) = x^k$.

Definition. Suppose that a subset $U \subset \Omega_p$ has no isolated points. A function $f: U \to \Omega_p$ is called <u>locally analytic</u> if for every $a \in U$ there exist r and a_i such that for all x in $D_a(r) \cap U$

$$f(x) = \sum_{i=0}^{\infty} a_i (x-a)^i.$$

It is easy to check that a locally analytic function f can be differentiated in the usual way:

$$f'(x) = \lim_{\varepsilon\to 0} \frac{f(x+\varepsilon)-f(x)}{\varepsilon} = \sum i a_i (x-a)^{i-1} \quad \text{for } x \in D_a(r) \cap U.$$

Lemma. <u>If</u> f <u>is locally analytic on</u> Z_p, <u>then the limit</u> $\lim\limits_{n\to\infty} p^{-n} \sum\limits_{0\le j<p^n} f(j)$ <u>exists.</u>

To prove the lemma, one easily reduces to the case when $f(x) = \Sigma\, a_i x^i$ on $D(1)$. (Thus, $a_i \to 0$.) Then we need to show that $\Sigma\, a_i B_i$ converges, but this follows because $|B_k|_p \le p$. (More precisely, we have already seen that $B_k \in Z_p$ if $p-1 \nmid k$, and one can similarly use the p-adic integral formula

$$(1 - p^{k-1})\, B_k = \frac{k}{1 - r^k} \int_{Z_p} x^{k-1}\, d\mu(x)$$

to show that $pB_k \in Z_p$ if $p-1|k$.)

The approach of Kubota-Leopoldt is based on this lemma and the formulas (7.1) and (7.2). In order to obtain a p-adic function out of (7.1) as k approaches some p-adic s, we must omit the j which are divisible by p and also restrict to $k \equiv k_0 \pmod{p-1}$ for some fixed k_0. If we write $j = \langle j \rangle \omega(j)$ for $p \nmid j$, then we obtain for $k \equiv k_0 \pmod{p-1}$ (see §3 for the definition of ζ^*):

$$\begin{aligned}
\zeta^*(1-k) &= (1-p^{k-1})\left(-\frac{B_k}{k}\right) \\
&= -\frac{1}{k} \lim_{n \to \infty} p^{-n}\left(\sum_{0 \le j < p^n} j^k - p^k \sum_{0 \le j < p^{n-1}} j^k\right) \\
&= -\frac{1}{k} \lim_{n \to \infty} p^{-n} \sum_{0 \le j < p^n,\ p \nmid j} \langle j \rangle^k\, \omega(j)^{k_0}.
\end{aligned}$$

We can now define $\zeta_{p,k_0}(s)$ by replacing k by $1-s$ and applying the lemma to $f(x) = \langle x \rangle^{1-s}\, \omega(x)^{k_0}$ (we take $f(x) = 0$ on pZ_p). Similarly, the p-adic L-function for a Dirichlet character χ: $(Z/dZ)^* \longrightarrow \Omega_p^*$ can be defined by setting

$$L_p(1-s,\chi) = -\frac{1}{s} \lim_{n \to \infty} \frac{1}{dp^n} \sum_{0 \le j < dp^n,\ p \nmid j} \langle j \rangle^s\, \chi(j). \qquad (7.3)$$

This approach to the construction of ζ_{p,k_0} and L_p can be generalized as follows. Let $X = \varprojlim_N (Z/dp^N Z)$, as in §2. We call a function $f(x,s)$ on $X \times U$ (where U is a subset of Ω_p with no isolated points) <u>locally analytic</u> if every $(a,b) \in X \times U$ has a neighborhood $(a + dp^N Z_p) \times (D_b(r) \cap U)$ on which $f(x,s) = \Sigma\, a_{ij}(x-a)^i(s-b)^j$. Then it is easy to show that

$$F(s) = \lim_{n\to\infty} \frac{1}{dp^n} \sum_{0\le j<dp^n} f(j,s)$$

exists and is a locally analytic function of $s \in U$ (see [22]). In the case of $L_p(s,\chi)$, we had $U = Z_p$ and

$$f(x,s) = \begin{cases} \langle x\rangle^s \chi(x) & \text{if } x \in X^*; \\ 0 & \text{otherwise.} \end{cases}$$

A special case of this construction leads to the p-adic log gamma function. We start with a function on $Z_p \times U$ of the form $f(x+s)$ with U now a subset of Ω_p which is invariant under translation by Z_p.

Lemma. *Suppose that* $f(x)$ *is locally analytic on* $s + Z_p$ *for some fixed* $s \in \Omega_p$. *Let*

$$F(s) = \lim_{n\to\infty} p^{-n} \sum_{0\le j<p^n} f(s+j).$$

Then F *is locally analytic on* $s + Z_p$, *and*

$$F(x+1) - F(x) = f'(x).$$

The proof is easy; the last assertion follows because

$$F(x+1) - F(x) = \lim (f(x+p^n) - f(x))/p^n.$$

The classical log gamma function satisfies $\log \Gamma(x+1) - \log \Gamma(x) = \log x$. So, by the lemma, the natural way to obtain a p-adic analog is to let $f'(x) = \ln_p x$, i.e., $f(x) = x \ln_p x - x$ (see the remark at the end of §I.3). Thus, J. Diamond [22] defined his p-adic log gamma function as

$$G_p(x) = \lim_{n\to\infty} p^{-n} \sum_{0\le j<p^n} (x+j)\ln_p(x+j) - (x+j) \tag{7.4}$$

for $x \in \Omega_p - Z_p$. Thus,

$$G_p(x+1) - G_p(x) = \ln_p x. \tag{7.5}$$

Note that it is inevitable that a continuous p-adic function satisfying (7.5) not be defined on Z_p. Namely, $\ln_p 0$, and hence either $G_p(1)$ or $G_p(0)$, is not defined. It then follows by induction that G_p cannot be defined either on the positive integers or the negative integers, both of which are dense in Z_p.

The other possible candidate for a p-adic log gamma function, namely $\ln_p\Gamma_p$, _is_ defined on Z_p, but it only satisfies (7.5) when $x \in Z_p^*$ (see property (1) of Γ_p).

The two functions G_p and $\ln_p\Gamma_p$ are related as follows. First note that $G_p(x) + G_p(1-x) = 0$, as follows immediately from (7.4) (after replacing j by p^n-1-j). I now claim: If $x \in Z_p$, then

$$\ln_p\Gamma_p(x) = \sum_{0\le i<p,\ p\nmid i+x} G_p\left(\frac{i+x}{p}\right), \tag{7.6}$$

i.e., we omit the one value of i for which $G_p((i+x)/p)$ is not defined because $(i+x)/p \in Z_p$. To prove (7.6), we note that both sides vanish when $x = 0$, since $G_p(i/p) + G_p((p-i)/p) = 0$; and both sides of (7.6) change by the same amount when x is replaced by $x + 1$, namely by $\ln_p x$ if $x \in Z_p^*$ and by 0 if $x \in pZ_p$. Since the nonnegative integers are dense in Z_p, we have (7.6) for all $x \in Z_p$.

We discover an interesting relationship between G_p and the zeta function if we expand G_p in powers of $1/x$ for x large. Suppose $|x|_p > 1$. We have

$$\begin{aligned}
G_p(x) &= \lim_{n\to\infty} p^{-n}\sum_{0\le j<p^n}(x+j)\ln_p x \;+ \\
&\qquad x\lim_{n\to\infty} p^{-n}\sum_{0\le j<p^n}\left(1+\frac{j}{x}\right)\left(-1+\ln_p\left(1+\frac{j}{x}\right)\right) \\
&= (x-\tfrac{1}{2})\ln_p x - x + x\lim_{n\to\infty} p^{-n}\sum_{0\le j<p^n}\sum_{k=1}^{\infty}(-1)^{k+1}\left(\frac{j}{x}\right)^{k+1}\left(\frac{1}{k}-\frac{1}{k+1}\right) \\
&= (x-\tfrac{1}{2})\ln_p x - x + \sum_{k=1}^{\infty}\frac{B_{k+1}}{k(k+1)}x^{-k} \qquad (7.7)\\
&= (x-\tfrac{1}{2})\ln_p x - x + \frac{1}{12x} - \frac{1}{360x^3} + \cdots,
\end{aligned}$$

where we used (7.1) and the fact that $B_k = 0$ for odd $k \ge 3$.

Hence,

$$G_p(x) = (x - \frac{1}{2}) \ln_p x - x - \sum_{k=1}^{\infty} \zeta(-k) \frac{x^{-k}}{k} .$$

Roughly speaking, one might expect that, since $\zeta(-k)$ is an integral of t^k, $G_p(x)$ is essentially $-\sum \int t^k \frac{x^{-k}}{k} d\mu(t) = \int \ln_p(1 - \frac{t}{x}) d\mu(t)$. We shall later look more carefully at this possibility (see §8 and the Appendix).

Remark. In the classical case, Stirling's formula

$$n! = \sqrt{2\pi n} \frac{n^n}{e^n} e^{\theta/12n}, \qquad 0 < \theta < 1,$$

gives

$$\log \frac{\Gamma(x)}{\sqrt{2\pi}} = (x - \frac{1}{2}) \log x - x + \frac{\theta}{12x} .$$

Thus, G_p is actually the analog of $\log(\Gamma(x)/\sqrt{2\pi})$. (From a number theoretic point of view it is often natural to normalize the gamma function by dividing by $\sqrt{2\pi}$. For example, $\Gamma(\frac{1}{2})/\sqrt{2\pi} = 1/\sqrt{2}$ is algebraic; also, the right side of the Gauss multiplication formula becomes simpler, see property (5) at the beginning of §6.) Note that in the classical case the series (7.7) is only an asymptotic series. We cannot simply evaluate (7.7) at $x \in C$, since it diverges for all x: $|B_k|$ grows roughly like $k!$, in contrast to $|B_k|_p$, which is bounded.

Finally, we note the following "distribution property" of G_p, which follows immediately from the definition (7.4) and the fact that $\ln_p p = 0$:

$$G_p(x) = \sum_{0 \le i < p} G_p\left(\frac{x + i}{p}\right) \quad \text{for} \quad x \in \Omega_p - Z_p. \tag{7.8}$$

8. A formula for $L_p'(0,\chi)$

The purpose of this section is to prove a formula for $L_p'(0,\chi)$ which is analogous to a classical formula of Lerch (see [97], p. 271):

$$L'(0,\chi) = B_{1,\chi} \log d + \sum_{0<a<d} \chi(a) \log \Gamma(a/d)$$
$$= -L(0,\chi) \log d + \sum_{0<a<d} \chi(a) \log \Gamma(a/d).$$

We start by defining a twisted version of G_p:

$$G_{p,z}(x) = \lim_{n\to\infty} \frac{1}{rp^n} \sum_{0\leq j<rp^n} z^j (x+j)(\ln_p(x+j) - 1), \tag{8.1}$$

where $z^r = 1$, $x \in \Omega_p - Z_p$. In particular, $G_{p,1} = G_p$. The following properties of $G_{p,z}$ are proved in the same way as the analogous properties of G_p.

Proposition. _The limit (8.1) exists for_ $x \in \Omega_p - Z_p$ _and satisfies_:

$$z\, G_{p,z}(x + 1) - G_{p,z}(x) = \ln_p x \quad \underline{\text{for}}\ x \notin Z_p; \tag{8.2}$$

$$G_{p,z}(x) = \sum_{i=0}^{p-1} z^i\, G_{p,z^p}\left(\frac{x+i}{p}\right) \quad \underline{\text{for}}\ x \notin Z_p; \tag{8.3}$$

$$G_{p,z}(x) = B_{1,z} \ln_p x + \sum_{k=1}^{\infty} \frac{(-1)^k}{k} x^{-k} L(-k,1,z) \tag{8.4}$$

for $|x|_p > 1$, _where_ $B_{1,z} = 1/(z-1) = -L(0,1,z)$. (See formula (1.4) with $k = 0$, f trivial, $d = 1$, $z = \varepsilon$; in (8.4) we are supposing that $z \neq 1$.)

We now give an expression for $G_{p,z}$ in terms of the measure μ_z. Here $z^r = 1$, $z \neq 1$, and μ_z is the measure on Z_p defined by $\mu_z(a + p^N Z_p) = z^a/(1-z^{p^N})$.

Proposition.

$$G_{p,z}(x) = -\int_{Z_p} \ln_p(x + t)\, d\mu_z(t) \quad \underline{\text{for}}\ x \in \Omega_p - Z_p. \tag{8.5}$$

Proof. Let $\tilde{G}_{p,z}(x)$ denote the function on the right in (8.5). Then for $|x|_p > 1$ we have

$$\tilde{G}_{p,z}(x) = -\int_{Z_p} \ln_p x\, d\mu_z(t) - \int_{Z_p} \ln_p\left(1 + \frac{t}{x}\right) d\mu_z(t)$$
$$= -\ln_p x\, \mu_z(Z_p) + \sum_{k\geq 1} \frac{1}{k}(-1)^k \int_{Z_p} t^k\, d\mu_z(t)\, x^{-k}$$

$$= -L(0,1,z)\ \ln_p x + \sum_{k\geq 1} \frac{(-1)^k}{k} x^{-k} L(-k,1,z)$$

by (2.5). Thus, by (8.4), $\tilde{G}_{p,z}(x) = G_{p,z}(x)$ for $|x|_p > 1$.

Now let $U_n = \{x \in \Omega_p \mid |x-j|_p > p^{-n}$ for all $j \in Z_p\}$. Then $\Omega_p - Z_p = \bigcup_{n=0}^{\infty} U_n$. We prove that $\tilde{G}_{p,z}(x) = G_{p,z}(x)$ for $x \in U_n$ by induction on n. We just proved this equality for $n = 0$. If we show that $\tilde{G}_{p,z}$, like $G_{p,z}$, satisfies (8.3), then the induction step will follow, since $x \in U_{n+1} \Rightarrow (x+i)/p \in U_n$ for $i = 0, 1, \ldots, p-1$. But the change of variables $u = pt + i$ gives

$$z^i \int_{Z_p} \ln_p\left(\frac{x+i}{p} + t\right) d\mu_{z^p}(t) = \int_{i+pZ_p} \ln_p(x+u)\, d\mu_z(u)$$

if we use the fact that $\ln_p p = 0$ and the definition of μ_z (as in (3.2)). The property (3.4) then follows for $\tilde{G}_{p,z}$, and the proposition is proved.

Remark. If we define the <u>convolution</u> g of f with μ by $g(x) = \int_{Z_p} f(x+t)\, d\mu(t)$ for $x \in \Omega_p$ such that f is continuous on $x + Z_p$, then it follows from (2.3) and the definition of μ_z that $zg(x+1) - g(x) = -f(x)$ when $\mu = \mu_z$. Thus, if we take the preceding proposition as the <u>definition</u> of $G_{p,z}$, then property (8.2) follows from this equality with $f = -\ln_p$.

Corollary.

$$G_{p,z}^{(k)}(x) = (-1)^k (k-1)! \int_{Z_p} \frac{d\mu_z(t)}{(x+t)^k} \quad \text{for } x \in \Omega_p - Z_p,\ k \geq 1. \qquad (8.6)$$

Theorem (Diamond [22] and Ferrero-Greenberg [29]). *Let χ be a nontrivial character of conductor d. Then*

$$L_p'(0,\chi) = \sum_{0<a<pd,\ p\nmid a} \chi_1(a)\, G_p\left(\frac{a}{pd}\right) - L_p(0,\chi) \ln_p d. \qquad (8.7)$$

Proof. In order to relate twisted and untwisted G_p, we need a lemma. As usual, r is any positive integer prime to pd.

Lemma. Let $0 < a < pd$, $p \nmid a$, $B_1(x) = x - \frac{1}{2}$, and define a', $0 < a' < pd$, by: $ra' \equiv a \pmod{pd}$. Then

$$\sum_{z^r=1} z^a G_{p,z^{pd}}\left(\frac{a}{pd}\right) = r\left(G_p\left(\frac{a'}{pd}\right) + \ln_p r \, B_1\left(\frac{a'}{pd}\right)\right). \tag{8.8}$$

To prove (8.8), we use (8.1) to write the left side as

$$\lim_{n\to\infty} \frac{1}{rp^n} \sum_{0\le j<rp^n} \sum_{z^r=1} z^{a+pdj}\left(\frac{a}{pd}+j\right)\left(\ln_p\left(\frac{a}{pd}+j\right)-1\right)$$

$$= r \lim_{n\to\infty} \frac{1}{rp^n} \sum_{\substack{0\le j<rp^n \\ j\equiv -a/pd \pmod r}} \left(\frac{a}{pd}+j\right)\left(\ln_p\left(\frac{a}{pd}+j\right)-1\right)$$

$$= \lim_{n\to\infty} p^{-n} \sum_{0\le j<p^n} \left(\frac{a}{pd}+a''+rj\right)\left(\ln_p\left(\frac{a}{pd}+a''+rj\right)-1\right),$$

where $0 < a'' < r$, $a'' \equiv -a/pd \pmod r$. Since $\frac{1}{r}\left(\frac{a}{pd}+a''\right) = \frac{a'}{pd}$, we find that the left side of (8.8) equals

$$r \lim_{n\to\infty} p^{-n} \sum_{0\le j<p^n} \left(\frac{a'}{pd}+j\right)\left(\ln_p r + \ln_p\left(\frac{a'}{pd}+j\right)-1\right)$$

$$= r\left(B_1\left(\frac{a'}{pd}\right)\ln_p r + G_p\left(\frac{a'}{pd}\right)\right).$$

We now proceed to the proof of the theorem. First, a twisted version of (8.7) (for $z \ne 1$) follows immediately by differentiating under the integral sign in the definition

$$L_p(s,\chi,z) = \int_{X^*} \langle x\rangle^{-s} \chi_1(x)\, d\mu_z(x)$$

and then setting $s = 0$. Namely, we have:

$$L_p'(0,\chi,z) = -\int_{X^*} \ln_p x\, \chi_1(x)\, d\mu_z(x)$$

$$= -\ln_p(pd)\int_{X^*}\chi_1 d\mu_z - \sum_{\substack{0<a<pd \\ p\nmid a}} \chi_1(a)\int_{a+pdZ_p} \ln_p\left(\frac{x}{pd}\right) d\mu_z(x)$$

$$= -L_p(0,\chi,z)\ln_p d - \sum_{\substack{0<a<pd \\ p\nmid a}} \chi_1(a) z^a \int_{Z_p} \ln_p\left(\frac{a}{pd}+t\right) d\mu_{z^{pd}}(x)$$

$$= -L_p(0,\chi,z)\ \ln_p d \ + \sum_{0<a<pd,\ p\nmid a} \chi_1(a) z^a\ G_{p,z^{pd}}\left(\frac{a}{pd}\right).$$

Now let A denote the right side of (8.7). We must show that $L'_p(0,\chi) = A$. Summing the twisted version of (8.7) over $z \neq 1$ with $z^r = 1$, and adding $A + L'_p(0,\chi)$ to both sides, we obtain

$$A + \sum_{z^r=1} L'_p(0,\chi,z) = L'_p(0,\chi) - \sum_{z^r=1} L_p(0,\chi,z)\ \ln_p d$$
$$+ \sum_{\substack{0<a<pd \\ p\nmid a}} \chi_1(a) \sum_{z^r=1} z^a G_{p,z^{pd}}\left(\frac{a}{pd}\right).$$

Note that the relation (4.5) between the twisted and untwisted L_p gives

$$\langle r\rangle^{1-s}\chi(r)\ L_p(s,\chi) = \sum_{z^r=1} L_p(s,\chi,z), \tag{8.9}$$

and, if we differentiate,

$$-\ln_p r\ \langle r\rangle^{1-s}\ \chi(r)\ L_p(s,\chi) + \langle r\rangle^{1-s}\ \chi(r)\ L'_p(s,\chi)$$
$$= \sum_{z^r=1} L'_p(s,\chi,z). \tag{8.10}$$

Using (8.9) with $s = 0$ and the lemma (8.8), we have

$$A + \sum_{z^r=1} L'_p(0,\chi,z) = L'_p(0,\chi) - \langle r\rangle\chi(r)\ L_p(0,\chi)\ \ln_p d$$
$$+ r \sum_{\substack{0<a<pd \\ p\nmid a}} \chi_1(a)\left(G_p\left(\frac{a'}{pd}\right) + \left(\frac{a'}{pd} - \frac{1}{2}\right)\ln_p r\right).$$

Note that $\langle r\rangle\chi(r) = r\chi_1(r)$ and $\chi_1(a') = \chi_1(a/r)$. Now using (8.10) with $s = 0$, we obtain

$$A + r\chi_1(r)L'_p(0,\chi) - r\chi_1(r)\ln_p r\ L_p(0,\chi) = L'_p(0,\chi) -$$
$$r\chi_1(r)L_p(0,\chi)\ln_p d + r\chi_1(r) \sum_{0<a'<pd,\ p\nmid a'} \chi_1(a')G_p\left(\frac{a'}{pd}\right)$$
$$+ r\chi_1(r)\ln_p r \sum_{0<a'<pd,\ p\nmid a'} \chi_1(a')\frac{a'}{pd}\ .$$

Since the last sum on the right equals $B_{1,\chi_1} = -L_p(0,\chi)$, we cancel that term and obtain

$$A + r\chi_1(r)\, L_p'(0,\chi) = L_p'(0,\chi) + r\chi_1(r)\, A,$$

using the definition of A. Since $r\chi_1(r) \neq 1$, this gives $L_p'(0,\chi) = A$, and the theorem is proved.

Corollary.

$$L_p'(0,\chi) = \sum_{0<a<d} \chi_1(a)\, \ln_p \Gamma_p\left(\frac{a}{d}\right) - L_p(0,\chi)\, \ln_p d. \qquad (8.11)$$

The corollary follows immediately by using the relation (7.6) between G_p and $\ln_p \Gamma_p$ in the formula (8.7).

Remark. In a very similar manner one can express the values of L_p at positive integers in terms of special values of the successive derivatives of $\ln_p \Gamma_p$ (see [23], [56]). If D^k denotes the k-th derivative, one has:

$$L_p(k,\chi_{k-1}) = \frac{(-d)^{-k}}{(k-1)!} \sum_{0<a<d} \chi(a) \left(D^k \ln_p \Gamma_p\right)\left(\frac{a}{d}\right) \quad \text{for } k \geq 1.$$

III. GAUSS SUMS AND THE p-ADIC GAMMA FUNCTION

1. Gauss and Jacobi sums

Let F_q be a finite field, let K be a field (such as C or Ω_p), let

$$\psi: F_q \longrightarrow K^*$$

be an additive character, i.e., a nontrivial homomorphism from the additive group of F_q to the multiplicative group K^*, and let

$$\chi: F_q^* \longrightarrow K^*$$

be a multiplicative character, i.e., a homomorphism from the multiplicative group F_q^* to K^*. (Warning: Characters χ on F_q^* should not be confused with the Dirichlet characters on $(Z/dZ)^*$ which were considered in Chapter II and were also denoted χ.) The Gauss sum (in K) of χ and ψ is defined as

$$g(\chi,\psi) = -\sum_{x\in F_q^*} \chi(x)\,\psi(x).$$

If χ_1 and χ_2 are two multiplicative characters of F_q, then the Jacobi sum of (χ_1,χ_2) is defined as

$$J(\chi_1,\chi_2) = -\sum_{x\in F_q,\ x\neq 0,1} \chi_1(x)\,\chi_2(1-x).$$

(The Gauss and Jacobi sums are usually defined without the minus sign before the summation, but this definition is more convenient for our purposes.)

The Gauss and Jacobi sums satisfy the following elementary properties:

(1) If χ is not the trivial character, and if $\overline{\chi} = \chi^{-1}$ denotes the conjugate character, then

$$g(\chi,\psi)\, g(\overline{\chi},\psi) = \chi(-1)\cdot q.$$

(2) If χ is nontrivial and $K = C$, then

$$|g(\chi,\psi)| = \sqrt{q}.$$

(3) If $\chi_1\chi_2$ is nontrivial, then

$$J(\chi_1,\chi_2) = \frac{g(\chi_1,\psi)\, g(\chi_2,\psi)}{g(\chi_1\chi_2,\psi)}. \tag{1.1}$$

Note that the expression in property (3) does not depend on the choice of (nontrivial) additive character ψ.

Remark. $g(\chi,\psi)$ is the analog for F_q of the gamma function on R. Namely,

$$\Gamma(s) = \int_0^\infty x^s e^{-x} \frac{dx}{x},$$

i.e., $\Gamma(s)$ is the "sum" over the (positive) multiplicative group of the field (i.e., the integral with respect to its Haar measure dx/x) of the product of a multiplicative character $x \longmapsto x^s$ and an additive character $x \longmapsto e^{-x}$. The analog of property (1) is: $\Gamma(s)\,\Gamma(1-s) = \frac{\pi}{\sin(\pi s)}$, in which π plays the role of q and $\sin(\pi s)$, which is essentially $e^{\pi is} = (-1)^s$, plays the role of $\chi(-1)$. $J(\chi_1,\chi_2)$ is the analog of the beta-function

$$B(r,s) = \int_0^1 x^{r-1} (1-x)^{s-1}\, dx = \frac{\Gamma(r)\,\Gamma(s)}{\Gamma(r+s)}. \tag{1.2}$$

As an illustration of this striking analogy, compare the proof of the expression (1.1) for the Jacobi sum in terms of Gauss sums with the proof of the formula (1.2) for the beta-function in terms of the gamma function. Both proofs are easy, so let's write them side-by-side:

$B(r,s)\,\Gamma(r+s) =$	$J(\chi_1,\chi_2)\, g(\chi_1\chi_2,\psi) =$
$\int_0^1 x^{r-1}(1-x)^{s-1}dx \int_0^\infty y^{r+s-1}e^{-y}dy$	$\sum_{x\in F_q} \chi_1(x)\chi_2(1-x) \sum_{y\in F_q} \chi_1\chi_2(y)\psi(y)$
$= \int_0^1\int_0^\infty (xy)^{r-1}[(1-x)y]^{s-1}e^{-y}\, y\,dx\,dy$	$= \sum_{x,y\in F_q} \chi_1(xy)\chi_2((1-x)y)\psi(y)$

change of variables

$$u = xy, \quad v = (1-x)y$$

$$x = \frac{u}{u+v}, \quad y = u+v$$

$= \int_0^\infty\int_0^\infty u^{r-1}v^{s-1}e^{-u-v}du\,dv$	$= \sum_{u,v\in F_q} \chi_1(u)\chi_2(v)\psi(u+v)$
$= \int_0^\infty u^{r-1}e^{-u}du \int_0^\infty v^{s-1}e^{-v}dv$	$= \sum_{u\in F_q} \chi_1(u)\psi(u) \sum_{v\in F_q} \chi_2(v)\psi(v)$
$= \Gamma(r)\,\Gamma(s).$	$= g(\chi_1,\psi)\, g(\chi_2,\psi)$

The analogy between Gauss and Jacobi sums and gamma and beta functions goes deeper. The purpose of this chapter is to show that Gauss sums are essentially values of the p-adic gamma function.

2. Fermat curves

Let K be a field containing d d-th roots of unity, for example, C or Ω_p or F_q when $q \equiv 1 \pmod d$. We let μ_d denote the set of d-th roots of unity. If K is of characteristic p, we must have $p \nmid d$. The projective Fermat curve $F(d)$, $d > 2$, is defined by $X^d + Y^d = Z^d$. The affine curve $F(d)^{aff}$ is defined by $x^d + y^d = 1$ $(x = X/Z,\ y = Y/Z)$. The group $\mu_d \times \mu_d$ operates on $F(d)$ and $F(d)^{aff}$ by

$$(\xi,\xi')(x,y) = (\xi x, \xi' y), \qquad \xi, \xi' \in \mu_d. \tag{2.1}$$

We first consider $F(d)$ over C and study the groups $H_1(F(d),Q)$

and $H^1_{DR}(F(d)/Q)$, and the action of $\mu_d \times \mu_d$ on these groups.

To describe the homology $H_1(F(d),Q)$, start with the path

$$\gamma_0\colon [0,1] \longrightarrow F(d)^{aff}, \quad \gamma_0(t) = (t, \sqrt[d]{1-t^d}).$$

Fix a primitive d-th root of unity ξ. Let

$$\gamma = \gamma_0 - (1,\xi)\gamma_0 + (\xi,\xi)\gamma_0 - (\xi,1)\gamma_0,$$

so that γ goes first from $(0,1)$ to $(1,0)$, then from $(1,0)$ to $(0,\xi)$, then from $(0,\xi)$ to $(\xi,0)$, then from $(\xi,0)$ to $(0,1)$.

Note that $F(d)$ is a nonsingular plane curve of degree d. From algebraic geometry we know the following

Fact. *The* $2g = (d-1)(d-2)$ *differential forms*

$$\omega_{r,s} = x^{r-1}y^{s-1}\frac{dx}{y^{d-1}}, \quad 1 \le r,\ s \le d-1, \quad r+s \ne d,$$

form a basis for

$$H^1_{DR}(F(d)/Q) = \frac{\{\text{differentials of the second kind}\}}{\{\text{exact differentials}\}},$$

and the g *forms with* $r+s < d$ *form a basis for the holomorphic forms in* $H^1_{DR}(F(d)/Q)$.

Note that for $(\xi,\xi') \in \mu_d \times \mu_d$

$$(\xi,\xi')^*\,\omega_{r,s} = \xi^{r-1}\xi'^{s-1}\frac{\xi}{\xi'^{d-1}}\,\omega_{r,s} = \xi^r\xi'^s\,\omega_{r,s},$$

so that $\omega_{r,s}$ is an eigen-form for $\mu_d \times \mu_d$, which acts by the character $\chi_{r,s}\colon (\xi,\xi') \longmapsto \xi^r\xi'^s$, i.e.,

$$\alpha^*\,\omega_{r,s} = \chi_{r,s}(\alpha)\,\omega_{r,s} \quad \text{for } \alpha \in \mu_d \times \mu_d.$$

Remark. The above assertion about $\omega_{r,s}$ with $r+s < d$ forming a basis for the holomorphic differentials is true for any nonsingular plane curve of degree d (see [87], p. 171-173). The assertion about differentials of the second kind can be proved as follows.

First note that the points at infinity $F(d) - F(d)^{aff}$ are given in the coordinates $u = Z/X = 1/x$, $v = Y/X = y/x$ by $u = 0$, $v = \zeta$, where ζ runs through the d d-th roots of -1. Consider the $(d-1)$-dimensional vector space of differentials on $F(d)$ of the form $H(x,y) \frac{dx}{y^{d-1}}$, where $H(x,y) = x^{d-2}P(y/x)$ is a homogeneous polynomial of degree $d-2$. In the (u,v)-coordinates,

$$H(x,y) \frac{dx}{y^{d-1}} = -x\ (x/y)^{d-1}P(y/x)\left(-\frac{dx}{x^2}\right) = -\frac{1}{u}P(v)\ \frac{du}{v^{d-1}},$$

which has residue $\mathrm{res}_\zeta = -\zeta^{1-d}P(\zeta) = \zeta P(\zeta) = \Sigma_{0<i<d}\ a_{i-1}\zeta^i$ at the point at infinity $(0,\zeta)$. The sum of the residues is clearly zero, but the residues cannot all be zero unless $H(x,y) = 0$, since $a_{i-1} = \frac{1}{d-1}\Sigma_\zeta\ \zeta^{-i}\mathrm{res}_\zeta$. Thus, the map from the $(d-1)$-dimensional vector space of H's to the $(d-1)$-dimensional vector space of possible residues at infinity whose sum is zero, is surjective, i.e., any differential form on $F(d)^{aff}$ differs from a differential of the second kind (i.e., one with all residues zero) by a differential of the form $H(x,y)dx/y^{d-1}$. So for suitable $H(x,y)$, homogeneous of degree $d-2$, we can write

$$\omega_{r,s} - H(x,y)\ \frac{dx}{y^{d-1}} = \text{a differential of the second kind;}$$

applying $(\xi,\xi)^*$ gives

$$\xi^{r+s}\omega_{r,s} - H(x,y)\ \frac{dx}{y^{d-1}} = \text{another differential of the 2nd kind.}$$

If $r+s \not\equiv 0 \pmod d$, we can subtract and conclude that $\omega_{r,s}$ is a differential of the second kind. Finally, to show that the $2g$ differentials $\omega_{r,s}$ with $r,\ s < d$ are linearly independent modulo exact differentials, it suffices to use the fact that they are not exact (see (2.2) below) and they are eigen-forms for $\mu_d \times \mu_d$ with distinct characters.

We can thus write

$$H^1_{DR}(F(d)/Q) = \bigoplus_{1\le r,s<d,\ r+s\ne d} H^1_{DR}(F(d)/Q)^{\chi_{r,s}},$$

where the space $H^1_{DR}(F(d)/Q)^{\chi_{r,s}}$ of forms on which $\mu_d \times \mu_d$ acts by $\chi_{r,s}$ is one-dimensional and is spanned by $\omega_{r,s}$.

After establishing these facts about $H^1_{DR}(F(d)/Q)$, it is not hard to show that the classes of the paths $\{(\xi,\xi')\gamma\}_{(\xi,\xi')\in\mu_d\times\mu_d}$ span the homology $H_1(F(d),Q)$.

We now compute:

$$\int_\gamma \omega_{r,s} = \int_{\gamma_0}\omega_{r,s} - \int_{(1,\xi)\gamma_0}\omega_{r,s} + \int_{(\xi,\xi)\gamma_0}\omega_{r,s} - \int_{(\xi,1)\gamma_0}\omega_{r,s}$$

$$= \int_{\gamma_0}\left(1 - (1,\xi)^* + (\xi,\xi)^* - (\xi,1)^*\right)\omega_{r,s}$$

$$= (1 - \xi^s + \xi^{r+s} - \xi^r)\int_{\gamma_0}\omega_{r,s}$$

$$= (1-\xi^r)(1-\xi^s)\int_{\gamma_0} x^{r-1}y^{s-1}\frac{dx}{y^{d-1}}\ .$$

But

$$\int_{\gamma_0} x^r y^{s-d}\frac{dx}{x} = \int_0^1 t^r(1-t^d)^{s/d-1}\frac{dt}{t}$$

$$= \frac{1}{d}\int_0^1 u^{r/d}\,(1-u)^{s/d-1}\frac{du}{u} \qquad (\text{here } u = t^d)$$

$$= \frac{1}{d}\,B\left(\frac{r}{d},\frac{s}{d}\right) \qquad (\text{see } (1.2)).$$

Thus

$$\int_\gamma \omega_{r,s} = \frac{(1-\xi^r)(1-\xi^s)}{d}\,B\left(\frac{r}{d},\frac{s}{d}\right). \tag{2.2}$$

3. L-series for algebraic varieties (not to be confused with Dirichlet L-series)

Let V_0 be a separable algebraic variety of finite type over F_q, and let V be obtained from V_0 by extending scalars to the algebraic closure $\overline{F}_q$: $V = V_0 \otimes \overline{F}_q$. The Frobenius map $F: V \to V$ is the map which raises coordinates to the q-th power (in terms of coordinate rings, it raises variables to the q-th power and keeps coefficients fixed). Thus, the F_q-points of V_0 are the fixed points of the Frobenius $\mathcal{F}$ in V.

If $g \in \text{End } V$, let $|\text{fix}(g)|$ denote the number of fixed points of g on V. Thus, $|V_0(F_{q^n})| = |\text{fix}(\mathcal{F}^n)|$.

Let G be a finite group of automorphisms of V_0 over F_q, let $\rho: G \to GL(W)$ be a finite dimensional representation of G in a vector space W over a field K of characteristic zero, and let $\chi = \text{Tr}(\rho)$. Then we define

$$L(V_0/F_q, G, \rho) = \exp\left(\sum_{n=1}^{\infty} \frac{T^n}{n} \frac{1}{|G|} \sum_{g \in G} \chi(g^{-1}) |\text{fix}(\mathcal{F}^n \circ g)|\right) \qquad (3.1)$$

$$\in K[[T]].$$

For example, if $G = \{1\}$, then this is merely

$$Z(V_0/F_q) = \exp\left(\sum_{n=1}^{\infty} \frac{T^n}{n} |V_0(F_{q^n})|\right).$$

We shall be interested in the case of Fermat curves. Let V_0 be the Zariski open subset of $F(d)$ where none of the coordinates vanish:

$$V_0 = \{(x,y) \mid x^d + y^d = 1, \ xy \neq 0\}.$$

As always, we suppose $q \equiv 1 \pmod d$, so that $\mu_d \subset F_q$. Let K be an extension of Q_p containing $Q_p(\sqrt[d]{1})$. We imbed μ_d in K by the Teichmüller map $x \mapsto \omega(x)$. Let $G = \mu_d \times \mu_d$, which acts on V_0 by (2.1). Let χ_1 and χ_2 be two characters of μ_d with values in K, i.e., χ_i is of the form $x \mapsto \omega(x)^{a_i}$ $(i = 1, 2)$. Let $\rho = \chi = \chi_1 \times \chi_2$. Let $\tilde{\chi}_i$ $(i = 1, 2)$ be the character on F_q^* given by $x \mapsto \chi_i\left(x^{(q-1)/d}\right)$.

Claim. <u>The coefficient of</u> T <u>in the exponent in (3.1) is</u>

equal to $-J(\tilde{\chi}_1,\tilde{\chi}_2)$.

To prove this claim, note that $(x,y) \in V$ is fixed by $F\circ(\xi,\xi')$ whenever $(\xi x^q,\xi' y^q) = (x,y)$. Thus, the inner sum in (3.1) for $n=1$ is

$$\sum_{\substack{(x,y)\in V \\ x^{q-1}\in\mu_d,\ y^{q-1}\in\mu_d}} \chi_1(x^{q-1})\ \chi_2(y^{q-1}).$$

But $x^{q-1}, y^{q-1} \in \mu_d$ if and only if $u=x^d$ and $v=y^d$ are in F_q^*. In that case $\chi_1(x^{q-1})\chi_2(y^{q-1}) = \tilde{\chi}_1(u)\tilde{\chi}_2(v)$. Since for each (u,v) there are $|G|$ pairs (x,y) with $u=x^d$, $v=y^d$, we obtain

$$\frac{1}{|G|}\sum_{g\in G} \chi(g^{-1})\ |\mathrm{fix}(F\circ g)| = \sum_{\substack{u,\,v\in F_q^* \\ u+v=1}} \tilde{\chi}_1(u)\tilde{\chi}_2(v)$$

$$= \sum_{u\in F_q^*,\ u\neq 0,1} \tilde{\chi}_1(u)\tilde{\chi}_2(1-u)$$

$$= -J(\tilde{\chi}_1,\tilde{\chi}_2).$$

We shall also be interested in the case of so-called "Artin-Schreier curves". Just as the Fermat curve is connected with Jacobi sums, similarly the Artin-Schreier curve

$$y^p - y = x^d, \quad p \nmid d, \tag{3.2}$$

is connected with Gauss sums. Let $A(d,p)$ denote the complete nonsingular model of the plane curve (3.2). Note that here p appears in the form of the equation. This is related to the fact that Gauss sums, unlike Jacobi sums, depend on an additive character ψ.

In the Gauss sums we study, the additive character ψ is always assumed to be of the form $\psi: F_q \xrightarrow{\mathrm{Tr}} F_p \xrightarrow{\psi_0} K^*$ for some character ψ_0 of F_p, i.e., it is obtained by pulling back an additive character of F_p by means of the trace map from F_q to F_p.

The curve $A(d,p)$ is a degree d covering of the y-line which is totally ramified over the p solutions of $y^p - y = 0$ and over the point at infinity, and is unramified elsewhere. It follows from the Hurwitz genus formula (see, e.g., [40], p. 301) that $2g - 2 = -2d + (p+1)(d-1)$, so that $2g = (d-1)(p-1)$.

Over a field of characteristic p containing μ_d, the curve $A(d,p)$ has two types of automorphisms:

$$x \longmapsto \xi x, \quad y \longmapsto y, \qquad \xi \in \mu_d; \tag{3.3}$$

and

$$x \longmapsto x, \quad y \longmapsto y + \alpha, \qquad \alpha \in F_p. \tag{3.4}$$

We shall see that the first type corresponds to the multiplicative character χ and the second type to the additive character ψ in the Gauss sum. Note that, as in the case of $F(d)$, the number $2g = (d-1)(p-1)$ is the number of pairs (χ,ψ) with both χ and $\psi = \psi_0 \circ \mathrm{Tr}$ nontrivial.

Let $V_0 = \{(x,y) \mid y^p - y = x^d, \ x \neq 0\}$, let $G = \mu_d \times Z/pZ$, and let $\rho = \chi\colon (\xi,\alpha) \longmapsto \chi_1(\xi)\psi_0(\alpha)$.

Claim. <u>The coefficient of</u> T <u>in the exponent of (3.1) is equal to</u> $-g(\tilde{\chi}_1,\psi)$.

This claim is proved in a manner similar to the previous one:

$$\sum_{g\in G} \chi(g^{-1}) |\mathrm{fix}(F\circ g)| = \sum_{\substack{(\xi,\alpha),\ (x,y) \\ y^p - y = x^d \neq 0 \\ (x,y) = (\xi x^q, y^q + \alpha)}} \chi^{-1}(\xi,\alpha)$$

$$= \sum_{\substack{(x,y) \\ y^p - y = x^d = u \in F_q^*}} \chi_1(u^{(q-1)/d})\ \psi_0(y^q - y)$$

$$= |G| \sum_{u \in F_q^*} \tilde{\chi}_1(u)\ \psi_0(\mathrm{Tr}_{F_q/F_p} u),$$

as desired.

We now look more closely at L-series of this type. In particular, we show that the "bad" points of $F(d)$ and $A(d,p)$ that were omitted in V_0 (the points where a coordinate is zero) can in fact be included without changing the L-series, if our characters are nontrivial.

We return to the general case of a variety V_0. The Frobenius F acts on the $\overline{F}_q$-points of V_0. A closed point x of V_0 is the same as an orbit of F; $\deg x$ is the number of points in the orbit (equivalently, the degree over F_q of the field containing the coordinates of the $\overline{F}_q$-points in the orbit).

Now suppose that V_0 is quasi-projective, and G is a finite group of automorphisms of V_0 over F_q (hence commute with F). Let $X_0 = V_0/G$ with the induced Frobenius endomorphism.

We first consider the case when G has no fixed points, i.e., for all $g \in G$ and all $v \in V_0(\overline{F}_q)$, if $g \neq 1$, then $gv \neq v$. To every closed point x_0 in X_0 of degree N, we associate a conjugacy class $\mathrm{Frob}(x_0)$ in G as follows. First choose an x in the orbit x_0 and a $v \in V_0(\overline{F}_q)$ lying over x. Then $F^N v$ also lies over x, and so equals gv for some unique $g \in G$. Changing our choice of x in the orbit x_0 or our choice of v lying over x only changes g by conjugation. Hence we obtain a conjugacy class $\mathrm{Frob}(x_0)$ in G depending only on x_0. (In our application later, G will be abelian, so that $\mathrm{Frob}(x_0)$ will be a well-defined element.)

Let ρ be a representation of G in a finite dimensional vector space W over a field K of characteristic zero, and let $\chi = \mathrm{Tr}\ \rho$. Then

Claim.

$$L(V_0/F_q, G, \rho) = \prod \frac{1}{\mathrm{Det}\left(1 - \rho(\mathrm{Frob}(x_0)) \cdot T^{\deg x_0}\right)},$$

<u>where the product is taken over all closed points</u> x_0 <u>of</u> X_0.

To prove this claim, we take the log of both sides and use the fact that $\log \operatorname{Det}(1-MT) = -\sum \frac{T^n}{n} \operatorname{Trace}(M^n)$ for any matrix M. Then we need only show that

$$\frac{1}{|G|} \sum_{g \in G} \chi(g^{-1}) \, |\operatorname{fix}(F^n g \text{ on } V)|$$

is equal to

$$n \cdot \text{coefficient of } T^n \text{ in } \sum_{x_0} \sum_{r=1}^{\infty} \frac{T^{r \deg x_0}}{r} \operatorname{Trace} \rho(\operatorname{Frob}(x_0))^r,$$

which can be written

$$\sum_{x_0 \text{ of degree } s|n} s \, \chi(\operatorname{Frob}^{n/s}(x_0)),$$

and this equality follows easily by writing the first sum as the sum of $\chi(g)$ over all g and v with $F^n v = gv$ and then using the definition of $\operatorname{Frob}(x_0)$.

Now let us allow G to have fixed points. For $v \in V_0(\overline{F}_q)$, let $I_v = \{g \in G \mid gv = v\}$, the "inertia group" of v. Let

$$W^{I_v} = \{w \in W \mid \rho(g) w = w \text{ for all } g \in I_v\}.$$

Now $\operatorname{Frob}(x_0)$ is only defined up to multiplication by elements of I_v, as well as conjugation; nevertheless, the determinant of $1 - T^{\deg x_0} \cdot \rho(\operatorname{Frob}(x_0))$ acting on W^{I_v} still depends only on x_0; and it is not hard to show that

$$L(V_0/F_q, G, \rho) = \prod_{x_0} \operatorname{Det}\left(1 - T^{\deg x_0} \rho(\operatorname{Frob}(x_0))\big|_{W^{I_v}}\right)^{-1}. \tag{3.5}$$

In our example $V_0 = F(d)$, $G = \mu_d \times \mu_d$, the fixed points occur

(1) when $Y = 0$, in which case $I_v = 1 \times \mu_d$;

(2) when $X = 0$, in which case $I_v = \mu_d \times 1$;

(3) when $Z = 0$, in which case I_v is the diagonal in $\mu_d \times \mu_d$.

If our character $\rho = \chi = \chi_r \times \chi_s$ has the property that χ_r, χ_s, and $\chi_r\chi_s$ are all nontrivial (i.e., $0 < r, s < d$, $r+s \neq d$), then in all cases I_v acts nontrivially on the one-dimensional space W. Hence $W^{I_v} = 0$, and there are no contributions to (3.5) from the points with zero X, Y or Z coordinate.

Similarly, when $V_0 = A(d,p)$, $G = \mu_d \times Z/pZ$, the fixed points are

(1) the point at infinity, where $I_v = G$;

(2) the points with $x = 0$, where $I_v = \mu_d \times \{0\}$.

Again there is no contribution to (3.5) when the characters are nontrivial.

We now return to the general case of a variety V_0. Let ρ_{reg} be the regular representation of G. We have

$$\text{Trace } \rho_{reg}(g) = \begin{cases} |G| & \text{if } g = 1; \\ 0 & \text{if } g \neq 1. \end{cases}$$

It is immediate from the definition that

$$L(V_0/F_q, G, \rho_{reg}) = Z(V_0/F_q) = \exp \sum \frac{T^n}{n} |V_0(F_{q^n})|;$$

and also

$$L(V_0/F_q, G, \rho_{trivial}) = Z(X_0/F_q) \qquad (\text{recall } X_0 = V_0/G).$$

Since trivially we have

$$L(V_0/F_q, G, \rho_1 \oplus \rho_2) = L(V_0/F_q, G, \rho_1) \cdot L(V_0/F_q, G, \rho_2),$$

it follows that the decomposition

$$\rho_{reg} = \oplus \rho^{\deg \rho},$$

where the summation is over all irreducible representations of G, gives a corresponding product decomposition of $Z(V_0/F_q)$:

$$Z(V_0/F_q) = Z(X_0/F_q) \prod L(V_0/F_q, G, \rho)^{\deg \rho}, \tag{3.6}$$

where the product is over all nontrivial irreducible representations

ρ.

Suppose V_0 is a projective, nonsingular, geometrically connected curve over F_q, of genus g. Then X_0, the quotient of V_0 by G, is also nonsingular, say of genus g'. We have

$$Z(V_0/F_q) = \frac{\text{polynomial in } Z[T] \text{ of degree } 2g}{(1-T)\,(1-qT)}.$$

Moreover, if ρ is irreducible and nontrivial, then

$L(V_0/F_q,G,\rho)$ is a polynomial in T.

This was proved for curves by Weil in the 1940's, but it is only a conjecture in the general case of higher dimensional V_0. It then follows that

$$\sum \deg \rho \ \deg L(V_0/F_q,G,\rho) = 2g - 2g',$$

where the summation is over all irreducible nontrivial representations ρ.

4. Cohomology

Let $Z_p^{unr} = O_{Q_p^{unr}} = \{x \in Q_p^{unr} \mid |x|_p \le 1\}$; thus, Z_p^{unr} is the ring extension of Z_p generated by all N-th roots of unity with $p \nmid N$.

Fact. *For every prime* ℓ *there exists a functor* H^1

$$\left\{\begin{array}{l}\text{projective nonsingular}\\ \text{geometrically connected}\\ \text{curves of genus } g\\ \text{over } \overline{F}_q\end{array}\right\} \longrightarrow \left\{\begin{array}{l}\text{free modules (of rank } 2g)\\ \text{over } \begin{cases} Z_\ell & \text{if } \ell \neq p\\ Z_p^{unr} & \text{if } \ell = p\end{cases}\end{array}\right.$$

(*namely*, $H^1_{\text{étale}}(V,Z_\ell)$ *for* $\ell \neq p$ *and* $H^1_{\text{crystalline}}(V/Z_p^{unr})$ *for* $\ell = p$), *such that if* $V = V_0 \otimes_{F_q} \overline{F}_q$ *with* V_0 *defined over* F_q, *and if* F *is the* q-*th power Frobenius endomorphism, then*

$$\text{Trace}\left(F^* \mid H^1(V)\right) = 1 + q - |V_0(F_q)|. \qquad (4.1)$$

This fact implies that

$$Z(V_0/F_q) = \exp \sum_{n=1}^{\infty} \frac{T^n}{n} |V_0(F_{q^n})|$$

$$= \exp \sum_{n=1}^{\infty} \frac{T^n}{n} (1 + q^n - \mathrm{Trace}(\mathcal{F}^{*n}|H^1(V)))$$

$$= \frac{1}{1 - T} \cdot \frac{1}{1 - qT} \mathrm{Det}(1 - T \cdot \mathcal{F}^*|H^1(V)).$$

Here the Det term is a polynomial which is clearly in $Z[T]$.

Similarly, for L-series one can construct for $g \in G$ a "twisted" variety V_0' defined over F_q such that $V_0' \otimes_{F_q} \overline{F}_q \simeq V_0 \otimes_{F_q} \overline{F}_q = V$, while the Frobenius for V_0' is $\mathcal{F} \circ g$ ($\mathcal{F}$ is the Frobenius for V_0). Thus, (4.1) implies that

$$\mathrm{Trace}((\mathcal{F}^n \circ g)^*|H^1(V)) = 1 + q^n - |\mathrm{fix}(\mathcal{F}^n \circ g)|. \tag{4.2}$$

Now let ρ be an absolutely irreducible representation of G in a vector space over a field K which we assume contains Z_ℓ or Z_p^{unr}. Then the subspace of $H^1(V) \otimes K$ on which G acts by ρ is

$$(H^1(V) \otimes K)^\rho = \left(\frac{1}{|G|} \sum_{g \in G} \chi(g^{-1})\, g\right) (H^1(V) \otimes K),$$

and, using (4.2) and (3.1), one easily shows that

$$L(V_0/F_q, G, \rho) = \mathrm{Det}\left(1 - T \cdot \mathcal{F}^*|(H^1(V) \otimes K)^\rho\right). \tag{4.3}$$

We now apply (3.6) and (4.3) to our examples $F(d)$ and $A(d,p)$.

It follows from (4.3) that the coefficient of T in $L(V_0/F_q, G, \rho)$ is $-\mathrm{Trace}\ \mathcal{F}^*|(H^1(V) \otimes K)^\rho$. In the case $V_0 = F(d)$, $G = \mu_d \times \mu_d$, $\rho = \chi_r \times \chi_s$ with $0 < r, s < d$, $r + s \neq d$, we showed that this coefficient is $-J(\tilde{\chi}_r, \tilde{\chi}_s) \neq 0$. Since the $2g$ spaces $\left(H^1(F(d)) \otimes K\right)^{\chi_r \times \chi_s}$ are nonzero, it follows that each of them is one-dimensional, and we have the direct sum decomposition

$$H^1(F(d))\otimes K = \bigoplus_{r,s} \left(H^1(F(d))\otimes K\right)^{\chi_r\times\chi_s}.$$

Moreover, the eigen-value of F^* on $\left(H^1(F(d))\otimes K\right)^{\chi_r\times\chi_s}$ is Trace $F^* = J(\tilde{\chi}_r,\tilde{\chi}_s)$. Then (3.6) (with $V_0 = F(d)$ and $X_0 = V_0/G$ = the projective line) gives

$$Z(F(d)/F_q) = \frac{1}{(1-T)(1-qT)} \prod_{\substack{1\le r,s<d \\ r+s\neq d}} (1 - J(\tilde{\chi}_r,\tilde{\chi}_s)T),$$

as Weil explained in his famous paper in 1949 [99].

Next, let $V_0 = A(d,p)$: $y^p - y = x^d$ over F_q, $q \equiv 1 \pmod d$; $G = \mu_d \times Z/pZ$. We have shown that each of the $2g = (d-1)(p-1)$ characters $\rho = \chi\times\psi_0$ with χ and ψ_0 nontrivial gives nonzero Trace $F^*|\left(H^1(A(d,p))\otimes K\right)^{\chi\times\psi_0} = g(\tilde{\chi},\ \psi_0\circ\mathrm{Tr})$. Thus, we again have the decomposition

$$H^1(A(d,p))\otimes K = \bigoplus_{\chi,\ \psi_0\ \text{nontrivial}} \left(H^1(A(d,p))\otimes K\right)^{\chi\times\psi_0}$$

into one-dimensional eigenspaces with F acting by $g(\tilde{\chi},\ \psi_0\circ\mathrm{Tr})$ on the $\chi\times\psi_0$-component. We conclude:

$$Z(A(d,p)/F_q) = \frac{1}{(1-T)(1-qT)} \prod_{\chi,\ \psi_0\ \text{nontrivial}} (1 - g(\tilde{\chi},\ \psi_0\circ\mathrm{Tr})T).$$

We obtain a number theoretic corollary if we replace F_q by F_{q^n} and replace $g(\tilde{\chi},\psi)$ by the corresponding Gauss sum over F_{q^n}:

$$g_{F_{q^n}}(\tilde{\chi},\psi) = -\sum_{x\in F^*_{q^n}} \tilde{\chi}(N_{F_{q^n}/F_q}\ x)\ \psi(\mathrm{Tr}_{F_{q^n}/F_q}\ x).$$

Namely, we have

$$g_{F_{q^n}}(\tilde{\chi},\ \psi_0\circ\mathrm{Tr}_{F_q/F_p}) = \text{action of the } q^n\text{-power Frobenius on } (H^1(A(d,p))\otimes K)^{\chi\times\psi_0}$$

$$= F^{*n} \left| \left(H^1(A(d,p)) \otimes K\right)^{\chi \times \psi_0} \right.$$

$$= \left(g_{F_q}(\tilde{\chi}, \psi_0 \circ \mathrm{Tr})\right)^n,$$

which is known as the Hasse-Davenport relation for Gauss sums.

5. p-adic cohomology

By explicitly constructing a p-adic H^1 for $A(d,p)$, we shall derive a p-adic expression for $g(\tilde{\chi}, \psi_0 \circ \mathrm{Tr}) = F^* | (H^1 \otimes K)^{\chi \times \psi_0}$, showing how special values of the p-adic gamma function arise as eigen-values of Frobenius. For simplicity, we shall assume $p > 2$.

A key role will be played by the function $e^{\pi(x - x^p)}$ considered in §I.3. As before, we denote $E_\pi(x) = e^{\pi(x - x^p)}$.

Proposition. <u>There is a one-to-one correspondence between (p-1)-th roots π of $-p$ and nontrivial additive characters ψ_0 of F_p such that</u>

$$\psi_0(1) = E_\pi(1) \equiv 1 + \pi \pmod{\pi^2}.$$

<u>We then have</u>

$$\psi_0(a) = E_\pi(\omega(a)) \quad \underline{\text{for}} \quad a \in F_p.$$

(<u>Recall that</u> $\omega(a)$ <u>denotes the Teichmüller representative</u>.)

Proof. For $x \in D(1)$ we have: $E_\pi(x)^p = e^{p\pi(x - x^p)} = \sum \frac{(p\pi)^i}{i!}(x - x^p)^i$. (Thus, because of the p in the exponent, we <u>can</u> evaluate $E_\pi(x)^p$ by first evaluating the exponent and <u>then</u> expanding.) Hence, if $x \in D(1)$ satisfies $x - x^p = 0$ -- in other words, if $x = \omega(a)$ for some $a \in F_p$ -- then $E_\pi(x)^p = 1$. Thus, each $E_\pi(\omega(a))$ is a p-th root of 1. It follows from the expansion $E_\pi(x) = 1 + \pi x + \frac{1}{2}\pi^2 x^2 + \dots$ that $E_\pi(\omega(a)) \equiv 1 + \pi a \pmod{\pi^2}$. The proposition now follows easily.

One similarly shows that if $\psi = \psi_0 \circ \mathrm{Tr}_{F_q/F_p}$ is the additive character on F_q and if $\beta \in F_q$, then

$$\psi(\beta) = \prod_{0 \le i < f} E_\pi(\omega(\beta^{p^i})),$$

where $q = p^f$ and π is the (p-1)-th root of $-p$ corresponding to ψ_0.

Let V be a nonsingular algebraic variety over a perfect field k of characteristic p. In our application V will be the curve $A(d,p)$ and k will be F_q. Let R be a complete discrete valuation ring with maximal ideal M_R and residue field k, and let K be its fraction field. For us, R will be the ring of integers in $Q_p(\sqrt[q-1]{1}, \pi)$, where $\pi^{p-1} = -p$.

Washnitzer and Monsky [75] constructed an explicit version of p-adic H^i which to a so-called "special affine open set" associates a K-vector space. If $U = \{U_j\}$ is a covering of V by special affine open sets, then the map $U_j \longmapsto H^i(U_j)$ defines a Zariski presheaf H^i of K-vector spaces on V, such that the cohomology $E_2^{p,q} = H^p(V, H^q) \Longrightarrow H^*(V)$ conjecturally abuts to a "good" p-adic cohomology (in the sense of §4). This has been proved in the case when V is a curve. We shall only need the Washnitzer-Monsky H^1 for curves.

Theorem (Washnitzer-Monsky). <u>The functor</u>

$$V \longmapsto H^1(V) \underset{\text{def}}{=} H^0(V, H^1) \quad \text{(global sections)}$$

<u>on complete nonsingular geometrically connected curves</u> V <u>satisfies the properties</u>:

$$\operatorname{rank} H^1(V) = 2\,\operatorname{genus}(V);$$

$$\operatorname{Trace}(F^* | H^1(V)) = q + 1 - |\operatorname{fix}(F)|,$$

<u>when</u> $V = V_0 \otimes_{F_q} k$ <u>for some</u> V_0 <u>defined over</u> F_q, <u>where</u> $F = F_{F_q}$

is the q-*th power Frobenius endomorphism of* V.

The basic type of ring which goes into the construction of H^i is $R\langle\langle x_1,\dots,x_N\rangle\rangle$, which is defined as

$$\{\Sigma A_w x^w \in R[[x_1,\dots,x_N]] \mid \text{for some real number } B \text{ and } \varepsilon > 0 \text{ we have } \operatorname{ord}_p A_w \geq \varepsilon|w|+B\},$$

where $w = (w_1,\dots,w_N)$, $x^w = x_1^{w_1}\cdots x_N^{w_N}$, $|w| = \Sigma w_i$. Equivalently,

$$R\langle\langle x_1,\dots,x_N\rangle\rangle = \{f \in R[[x_1,\dots,x_N]] \mid \exists\, r > 1,\ f \text{ converges for } x_1,\dots,x_N \in D(r)\}.$$

It turns out that H^1 for the p-adic "affine line" will be $R\langle\langle x\rangle\rangle \otimes K \,/\, \frac{d}{dx} R\langle\langle x\rangle\rangle \otimes K = 0$, as expected.

Remark. In p-adic analysis $D(1)$ can often be thought of as the affine line, since it is the smallest disc containing the Teichmüller representatives of the points of the affine line over $\overline{F}_p$. One might ask why the simpler ring $A = \{f \in K[[x]] \mid f$ converges on $D(1)\}$ cannot be used. In fact, d/dx is not surjective on A. (For example, $a = \sum p^j x^{p^j - 1} = \frac{d}{dx}\sum x^{p^j} \in A$, but $\sum x^{p^j} \notin A$, so a is not $\frac{d}{dx}$ of any element of A.) In fact, $\operatorname{rank}(A/\frac{d}{dx}A) = \infty$ rather than $2\cdot\operatorname{genus}(\text{line}) = 0$. However, $\frac{d}{dx}$ *is* surjective on $R\langle\langle x\rangle\rangle \otimes K$.

For simplicity, suppose that the nonsingular curve V^{aff} is given by one equation $f(x,y) = 0$ (as will be the case in our application to $A(d,p)$). Consider the coordinate ring of the Zariski open subset U of V^{aff} where the tangent to the curve is not vertical:

$$k[x,y,t]/(f(x,y),\ t\frac{\partial f}{\partial y}-1).$$

Such an affine open set is called a "special affine" open set U over k. We now define the "dagger ring" for U to be the quotient

$$A^\dagger(U) = R\langle\langle x,y,t\rangle\rangle/(F(x,y),\ t\frac{\partial F}{\partial y} - 1),$$

where $F(x,y)$ is any fixed polynomial in $R[x,y]$ whose reduction modulo M_R is $f(x,y)$. Note that $A^\dagger(U)/M_R A^\dagger(U) = k[x,y,t]/(f,\ t\frac{\partial f}{\partial y} - 1)$, i.e., $A^\dagger(U)$ "lifts" the coordinate ring of the special affine open set U. (Of course, $A^\dagger(U)$ is not unique, since $F(x,y)$ is not unique.)

In our example $V = A(d,p)$, where $f(x,y) = y^p - y - x^d$, we have $\frac{\partial f}{\partial y} = -1$. We can take $F(x,y) = y^p - y - x^d$, so that $\partial F/\partial y = py^{p-1} - 1$, which is invertible in the ring $R\langle\langle y\rangle\rangle$ (since $\frac{1}{1 - py^{p-1}} = \sum p^j y^{(p-1)j}$). Thus, for $U = A(d,p)^{aff} = A(d,p)$ - {point at infinity}, we have

$$A^\dagger(U) = R\langle\langle x,y\rangle\rangle/(y^p - y - x^d).$$

Because $U = V$ - {point}, we are in an especially convenient situation for computing $H^1(V)$. Namely, if V is a complete, nonsingular geometrically connected curve over k such that V - {point} is a special affine U, then

$$H^1(V) \simeq H^1(V - \{\text{point}\}).$$

Roughly speaking, this is because, if $v_1,\ldots,v_s$ are finitely many points of V, then

$$H^1(V) \hookrightarrow H^1(V - \{v_i\})$$

can be identified as the subset whose residue at each point vanishes. Since the sum of the residues vanishes in $H^1(V - \{v_i\})$, this subset is all of $H^1(V - \{v_i\})$ if $s = 1$.

We now state a general fact (see [75]) about lifting a k-morphism $\phi_0\colon U_1 \longrightarrow U_2$ of special affine open sets to a morphism $\phi\colon A^\dagger(U_2) \longrightarrow A^\dagger(U_1)$ of their dagger rings. For simplicity of notation, we suppose that there is only one x variable, as will be the case in our application. Suppose that

$$U_i = \operatorname{Spec} k[x,y,t]/(f_i(x,y),\ t\frac{\partial f_i}{\partial y} - 1), \quad i = 1, 2.$$

First of all, such a lifting ϕ exists (ϕ is <u>not</u> unique). Next, suppose that we fix an element $\phi(x) \in A^{\dagger}(U_1)$ which reduces to $\phi_0(x)$ modulo M_R. Then there exists a unique choice of $\phi(y)$ and $\phi(t)$ lifting $\phi_0(y)$, $\phi_0(t)$. Finally, let

$$\Omega^{\bullet}_{A^{\dagger}(U)} = A^{\dagger}(U) \oplus A^{\dagger}(U)dx$$

with differential d, and let $H^1(U) = H^1(\Omega^{\bullet}_{A^{\dagger}(U)} \otimes K) = A^{\dagger}(U) \otimes K\, dx / d(A^{\dagger}(U) \otimes K)$. Then $H^1(U_i)$ is independent of the choice of F_i lifting f_i, and $\phi^*: H^1(U_2) \longrightarrow H^1(U_1)$ is independent of the choice of ϕ lifting ϕ_0. Thus, we may choose any convenient lifting.

We show how to choose a convenient lifting of ϕ_0 in the case $U = U_1 = U_2 = A(d,p)^{aff}$ and $\phi_0: x \longmapsto x,\ y \longmapsto y + \alpha$, $\alpha \in \{0,1,\ldots,p-1\}$. We want to construct an endomorphism ϕ of

$$A^{\dagger}(U) = R\langle\langle x,y\rangle\rangle/(y^p - y - x^d)$$

which reduces to ϕ_0 modulo M_R. It is simplest to choose $\phi(x) = x$. Then the above fact asserts that there exists a unique $\phi(y) \in A^{\dagger}(U)$ such that $\phi(y) \equiv y + \alpha \mod M_R$. To see this concretely, we note that $z = \phi(y) - y - \alpha$ must satisfy

$$(z + y + \alpha)^p - (z + y + \alpha) = y^p - y,$$

in other words,

$$z^p + \sum_{i=1}^{p-2} \binom{p}{i} (y+\alpha)^i z^{p-i} + (p(y+\alpha)^{p-1} - 1)\, z + (y+\alpha)^p - y^p - \alpha = 0. \quad (5.1)$$

See example (2) in §I.4.b, where we saw that if $\mathrm{ord}_p y > -\frac{1}{p-1}$, so that $\lambda = \mathrm{ord}_p((y+\alpha)^p - y^p - \alpha) > 0$, then there is a unique solution z with $0 < \mathrm{ord}_p z = \lambda$. This z can be expressed as a power series in y by first solving (5.1) mod $p^{2\lambda}$:

$$(p(y+\alpha)^{p-1} - 1)\, z + (y+\alpha)^p - y^p - \alpha = 0,$$

then substituting the approximate solution in place of all higher powers $z^2,\ldots,z^p$ in (5.1) and again solving the resulting linear equation for z, and so on. The result is a power series in y with coefficients in $Z_p \subset R$ which converges for $\mathrm{ord}_p y > -\frac{1}{p-1}$. Hence, $\phi(y) = z + y + \alpha \in R\langle\langle y\rangle\rangle \subset A^{\dagger}(U)$ is the desired lifting of $\phi_0(y) = y + \alpha$.

Of course, the other type of automorphism $\phi_0\colon x \longmapsto \xi x$, $y \longmapsto y$, where $\xi^d = 1$, $\xi \in F_q$, can be lifted simply to $\phi\colon x \longmapsto \omega(\xi)x$, $y \longmapsto y$, where $\omega(\xi) \in R$ is the Teichmüller representative.

Thus, the group $G = \mu_d \times Z/pZ$ acts on $A^{\dagger}(U)$, and hence on the dagger cohomology

$$H^1(A(d,p)) = A^{\dagger}(U)\otimes K\, dx \,/\, d(A^{\dagger}(U)\otimes K).$$

6. p-adic formula for Gauss sums

Because we claimed that the dagger H^1 is a good cohomology, the subspace of $H^1(A(d,p))$ on which $G = \mu_d \times Z/pZ$ acts by $\chi \times \psi_0$ should be one-dimensional (if χ and ψ_0 are nontrivial), and the q-th power Frobenius should act on it by $g(\tilde{\chi}, \psi_0 \circ \mathrm{Tr})$, where

$$\tilde{\chi}\colon\ F_q^* \xrightarrow{\ \frac{q-1}{d}\text{- th power}\ } \mu_d \xrightarrow{\ \chi\ } K^*.$$

Since G acts on $A^{+}(U) = R\langle\langle x,y\rangle\rangle/(y^{p}-y-x^{d})$, we can decompose

$$A^{+}(U)\otimes K = \bigoplus_{\text{all } \chi,\ \psi_0} \left(A^{+}(U)\otimes K\right)^{\chi\times\psi_0}.$$

The following proposition is due to Monsky. We let χ_a denote $\xi \longmapsto \omega(\xi)^{\frac{a}{d}(q-1)}$, we let ψ_{triv} denote the trivial character on F_p, and we let ψ_π denote the character $a \longmapsto E_\pi(\omega(a))$ on F_p (see the beginning of §5).

Proposition. <u>The subspace of</u> $A^{+}(U)$ <u>invariant under</u> $\{1\}\times Z/pZ \subset G$ <u>is</u> $R\langle\langle x\rangle\rangle$; <u>the subspace of</u> $A^{+}(U)\otimes K$ <u>on which</u> Z/pZ <u>acts by</u> ψ_π <u>is</u> $E_\pi(y)\left(R\langle\langle x\rangle\rangle\otimes K\right)$; $A^{+}(U)^{\chi_a\times\psi_{\text{triv}}} = x^{a}\, R\langle\langle x^{d}\rangle\rangle$; <u>and</u> $\left(A^{+}(U)\otimes K\right)^{\chi_a\times\psi_\pi} = E_\pi(y)\, x^{a}\left(R\langle\langle x^{d}\rangle\rangle\otimes K\right)$.

To prove this, first note that $E_\pi(y)$ and $E_\pi(-y) = 1/E_\pi(y)$ have coefficients in R (see §I.3) and converge on a disc strictly larger than $D(1)$; hence $E_\pi(y)$ is a unit in $R\langle\langle y\rangle\rangle$. In addition, $E_\pi(y)$ transforms by ψ_π under the action of Z/pZ, i.e.,

$$\phi_\alpha(E_\pi(y)) = \psi_\pi(\alpha)\, E_\pi(y), \quad \alpha\in F_p, \tag{6.1}$$

where ϕ_α is the lifting of $y \longmapsto y+\alpha$ constructed in §5. To see this, note that $\phi_\alpha(E_\pi(y))/E_\pi(y)$ is a p-th root of unity (independent of y), because

$$\left(E_\pi(\phi_\alpha(y))/E_\pi(y)\right)^{p} = e^{p\pi(\phi_\alpha(y)-\phi_\alpha(y)^{p})}\big/ e^{p\pi(y-y^{p})} = 1,$$

since $\phi_\alpha(y)^{p}-\phi_\alpha(y) = y^{p}-y$. (Note that the extra p in the exponent allows us to evaluate the exponent first, as at the beginning of §5.) To determine which p-th root, we set $y=0$:

$$E_\pi(\phi_\alpha(0))/E_\pi(0) = E_\pi(\phi_\alpha(0)) = E_\pi(\omega(\alpha)) = \psi_\pi(\alpha),$$

and (6.1) is proved.

Since $A^{+}(U)\otimes K$ clearly has rank p over $R\langle\langle x\rangle\rangle\otimes K$ (it equals $\bigoplus_{0\leq i<p} y^{i}R\langle\langle x\rangle\rangle\otimes K$), and the subspace on which Z/pZ acts by ψ_{π} (resp. ψ_{triv}) includes $E_{\pi}(y)\,R\langle\langle x\rangle\rangle\otimes K$ (resp. $R\langle\langle x\rangle\rangle\otimes K$), the first two assertions of the proposition follow. The assertions about the action of χ_a are obvious.

Remarks. 1. For fixed π, $\pi^{p-1} = -p$, we could also use the units $E_{\pi}(y)^{i} \in R\langle\langle y\rangle\rangle$, $i = 0, 1, \ldots, p-1$, since $E_{\pi}(y)^{i}$ transforms by $\psi_{\pi}^{\,i}$ under Z/pZ. Hence,

$$A^{+}(U)\otimes K = \bigoplus_{0\leq i<p-1} E_{\pi}(y)^{i}(R\langle\langle x\rangle\rangle\otimes K).$$

Note that $E_{\pi}(y)^{p} = e^{p\pi(y-y^{p})} = e^{-p\pi x^{d}} \in R\langle\langle x\rangle\rangle$. Thus, $E_{\pi}(y)$ is a Kummer generator of $A^{+}(U)\otimes K$ over $R\langle\langle x\rangle\rangle\otimes K$.

2. All of this applies to any curve of the form $y^{p}-y = f(x) \in R[x]$; the specific polynomial $f(x) = x^{d}$ was not needed in analyzing the action of Z/pZ.

Proposition. <u>For</u> $\psi_0 = \psi_{\pi}$ <u>and for</u> $\chi = \chi_a$, $1\leq a<d$, $H^{1}(A^{+}(U)\otimes K)^{\chi\times\psi_0}$ <u>is a one-dimensional</u> K-<u>vector space with basis</u> $E_{\pi}(y)\,x^{a}\frac{dx}{x}$; <u>if either</u> χ <u>or</u> ψ_0 <u>is trivial, this space is zero.</u>

Proof. We compute: $d\,E_{\pi}(y) = E_{\pi}(y)\,d\log E_{\pi}(y) = E_{\pi}(y)\,d(\pi(y-y^{p})) = \pi\,E_{\pi}(y)\,d(-x^{d}) = -\pi\,dx^{d-1}E_{\pi}(y)dx$. Thus,

$$d\left(E_{\pi}(y)x^{a}f(x^{d})\right) = E_{\pi}(y)x^{a}\frac{dx}{x}\left(-\pi dx^{d}f(x^{d}) + af(x^{d}) + dx^{d}f'(x^{d})\right),$$

so that for $a\geq 1$,

$$\left(H^1(A^\dagger(U)\otimes K)\right)^{\chi\times\psi_0} = \frac{E_\pi(y)\,x^a\,R\langle\langle x^d\rangle\rangle\frac{dx}{x}\otimes K}{d\left(E_\pi(y)\,x^a\,R\langle\langle x^d\rangle\rangle\otimes K\right)}$$

$$\simeq \frac{R\langle\langle x\rangle\rangle\otimes K}{\left(-\pi x+\frac{a}{d}+x\frac{d}{dx}\right)\left(R\langle\langle x\rangle\rangle\otimes K\right)}\,.$$

The proposition now asserts that the cokernel of $\partial = x\frac{d}{dx}+\frac{a}{d}-\pi x$ on $R\langle\langle x\rangle\rangle\otimes K$ is K if $a \geq 1$. (The triviality of $H^1(A^\dagger(U)\otimes K)^{\chi\times\psi_{triv}}$ is immediate, since $\frac{d}{dx}$ is surjective on $R\langle\langle x\rangle\rangle\otimes K$; the triviality of $H^1(A^\dagger(U)\otimes K)^{\chi_{triv}\times\psi_0}$ will be shown later.)

We compute:

$$x^{m+1} = \frac{m+\frac{a}{d}}{\pi}\,x^m + \partial\left(-\frac{x^m}{\pi}\right),$$

so that

$$\sum_{m=0}^{\infty} b_{m+1}x^{m+1} = \sum_{m=0}^{\infty} b_{m+1}\frac{\left(m+\frac{a}{d}\right)\left(m+\frac{a}{d}-1\right)\cdots\left(\frac{a}{d}\right)}{\pi^{m+1}}$$

$$-\,\partial\left(\sum_{n=1}^{\infty}x^n\sum_{m\geq n} b_{m+1}\frac{\left(m+\frac{a}{d}\right)\left(m+\frac{a}{d}-1\right)\cdots\left(n+1+\frac{a}{d}\right)}{\pi^{m+1-n}}\right). \tag{6.2}$$

Because we have

$$\operatorname{ord}_p b_{m+1}\frac{(m+1)!}{\pi^{m+1}}\binom{m+\frac{a}{d}}{m+1} \geq \operatorname{ord}_p b_{m+1}\frac{(m+1)!}{\pi^{m+1}} \quad \text{(see §I.3, formula (3.3))}$$

$$= \operatorname{ord}_p b_{m+1} - \frac{S_{m+1}}{p-1} \quad \text{(see §I.3, (3.1))},$$

while $\operatorname{ord}_p b_{m+1} \geq (m+1)\varepsilon + B$ when $\Sigma\, b_{m+1}x^{m+1} \in R\langle\langle x\rangle\rangle\otimes K$, it follows that the constant sum in (6.2) converges (since $S_{m+1} \leq (p-1)(\log_p m + 1)$). Similarly, ord_p of the inner sum inside the

∂ is at least

$$\min_{m \geq n} \left((m+1)\varepsilon + B - \frac{1+S_{m-n}}{p-1} \right) \geq n\varepsilon' + B'$$

for some $\varepsilon' > 0$ and some real number B'. Thus, any element f in $R\langle\langle x\rangle\rangle \otimes K$ can be written as $\text{const} + \partial g$, $g \in R\langle\langle x\rangle\rangle \otimes K$. It remains to show that 1 cannot be written as ∂g, $g \in R\langle\langle x\rangle\rangle \otimes K$, if $a \geq 1$, in which case we will have proved that the cokernel is precisely the constants. Suppose $g = \sum b_n x^n$, $1 = \partial g$. Then

$$1 + \pi x \sum b_n x^n = \left(x\frac{d}{dx} + \frac{a}{d} \right) \sum b_n x^n,$$

and comparing coefficients gives

$$1 = \frac{a}{d} b_0$$

$$b_n = \frac{\pi}{n + \frac{a}{d}} b_{n-1} \qquad (n \geq 1)$$

$$= \cdots = \frac{\pi^n}{(n+1)! \binom{n + \frac{a}{d}}{n+1}}.$$

Thus, $\operatorname{ord}_p b_n \leq (S_{n+1} - 1)/(p-1)$, which is <u>not</u> $\geq \varepsilon n + B$ for any $\varepsilon > 0$ and real number B. Thus, $g \notin R\langle\langle x\rangle\rangle \otimes K$.

The final assertion is that the cohomology is zero when $a = 0$. Roughly speaking, this is because the basis $E_\pi(y)\, x^a \frac{dx}{x}$ for the cokernel of ∂ only makes sense when $a \geq 1$. More precisely,

$$H^1(A^\dagger(U) \otimes K)^{\chi_{\text{triv}} \times \psi_\pi} = \frac{E_\pi(y)\; x^{d-1}\; R\langle\langle x^d\rangle\rangle dx \otimes K}{d\left(E_\pi(y)\; R\langle\langle x^d\rangle\rangle \otimes K\right)}$$

$$\simeq \frac{R\langle\langle x\rangle\rangle \otimes K}{\left(\frac{d}{dx} - \pi\right)(R\langle\langle x\rangle\rangle \otimes K)}.$$

To show that $\frac{d}{dx} - \pi$ is surjective on $R\langle\langle x\rangle\rangle \otimes K$, it suffices to show that its formal inverse $-\frac{1}{\pi}\left(1 - \frac{1}{\pi}\frac{d}{dx}\right)^{-1} = -\sum_{n=0}^{\infty} \pi^{-n-1}\left(\frac{d}{dx}\right)^n$ takes a series in $R\langle\langle x\rangle\rangle \otimes K$ to a series in $R\langle\langle x\rangle\rangle \otimes K$. But this is easy using the same estimate for $\frac{n!}{\pi^n}\binom{\alpha}{n}$ as above. This completes the proof of the proposition.

Recall that our Gauss sum $g(\tilde{\chi},\psi)$, $\psi = \psi_0 \circ \mathrm{Tr}$, is the action of the q-th power Frobenius $\mathcal{F} = \mathcal{F}_{\mathbb{F}_q}$ on $H^1(A^{\dagger}(U) \otimes K)^{\chi \times \psi_0}$. Since the curve $A(d,p)$ is defined over $\mathbb{F}_p$, the p-th power Frobenius $\mathcal{F}_0 = \mathcal{F}_{\mathbb{F}_p}$ also acts on $A(d,p)$ and hence on $H^1(A^{\dagger}(U) \otimes K)$; and we have $\mathcal{F} = \mathcal{F}_0^f$, where $q = p^f \equiv 1 \pmod d$. However, $\mathcal{F}_0$ does not commute with the action of $G = \mu_d \times \mathbb{Z}/p\mathbb{Z}$; its matrix in the eigen-vectors of G is not diagonal. More precisely, we have

Proposition. $\mathcal{F}_0^*\left(H^1(A^{\dagger}(U) \otimes K)^{\chi \times \psi_0}\right) \subset H^1(A^{\dagger}(U) \otimes K)^{\chi^p \times \psi_0}$.

This is an immediate consequence of the commutation relation

$$\mathcal{F}_0 \circ (\xi,\alpha)\ (x,y) = (\xi^p x^p,\ y^p + \alpha^p) = (\xi^p x^p,\ y^p + \alpha) = (\xi^p,\alpha) \circ \mathcal{F}_0(x,y),$$

for $(\xi,\alpha) \in G$.

Thus, if $\chi = \chi_a$, $1 \le a < d$, and if we let $a', a'', \ldots, a^{(j)}, \ldots, a^{(f-1)}$ denote the least positive residue of $p^j a \bmod d$, then for some constant $\lambda = \lambda(a,d,\pi)$ we have

$$\mathcal{F}_0^*\left(E_\pi(y)\, x^a \frac{dx}{x}\right) = \lambda\, E_\pi(y)\, x^{a'} \frac{dx}{x} \quad \text{in } H^1(A^{\dagger}(U) \otimes K). \tag{6.3}$$

Then we obtain

$$g(\tilde{\chi}_a,\ \psi_\pi \circ \mathrm{Tr}) = \prod_{j=0}^{f-1} \lambda(a^{(j)},d,\pi). \tag{6.4}$$

Proposition. $\lambda(a,d,\pi) = -p\,\pi^{-\frac{pa-a'}{d}}\,\Gamma_p\left(1-\frac{a'}{d}\right).$

This proposition, together with (6.4) and the elementary fact that $\frac{p-1}{d}\sum_{j=0}^{f-1} a^{(j)}$ = the sum of the p-adic digits in $a\frac{q-1}{d}$, imply:

Theorem.
$$g(\tilde{\chi}_a, \psi_\pi\circ\mathrm{Tr}) = \frac{(-p)^f \prod_{j=0}^{f-1} \Gamma_p\left(1-\left\langle\frac{p^j a}{d}\right\rangle\right)}{\pi^{S_{a(q-1)/d}}},$$

where $< >$ denotes the least positive residue modulo 1 (not to be confused with our earlier meanings of $< >$).

Proof of proposition. Let R be the ring of integers in $Q_p(\sqrt[d]{1},\pi)$. First we imbed

$$A^\dagger(U) = R\langle\langle x,y\rangle\rangle/(y^p - y - x^d) \overset{\phi}{\hookrightarrow} R[[x]] \qquad (6.5)$$

by sending y to the formal power series solution of $y^p - y = x^d$ near (0,0), i.e., with zero constant term: $y = -x^d + \ldots$. To see that such a solution exists (and is unique), one can use a version of Hensel's lemma (§I.2) for the "x-adic topology" in R[[x]] (the topology which says that two series are close together if their difference is divisible by a large power of x). Then $y=0$ is a solution mod x of $f(y) = y^p - y - x^d = 0$; moreover $f'(0) = -1 \neq 0$, so the existence and uniqueness of the desired series $\phi_y(x) \in R[[x]]$ is assured.

The imbedding (6.5) induces a homomorphism

$$H^1(A^\dagger(U)\otimes K) \longrightarrow H^1(R[[x]]\otimes K)$$

which commutes with the p-th power Frobenius F_0^*, where F_0 acts on $R[[x]]$ by sending x to x^p and acts on $A^\dagger(U)$ by sending x to x^p and y to the unique element which lifts y^p and satisfies $F_0(y)^p - F_0(y) = x^{pd}$.

We have

$$\phi\left(E_\pi(y)\, x^a \frac{dx}{x}\right) = \phi\left(e^{\pi(y-y^p)}\, x^a \frac{dx}{x}\right) = e^{-\pi x^d}\, x^a \frac{dx}{x}.$$

Thus, by (6.3),

$$F_0^*\left(e^{-\pi x^d}\, x^a \frac{dx}{x}\right) = \lambda(a,d,\pi)\, e^{-\pi x^d}\, x^{a'} \frac{dx}{x} \tag{6.6}$$

in $H^1(R[[x]]\otimes K)$. In other words, the difference between the two sides of this equality lie in $d(R[[x]]\otimes K)$. Note that

$$\sum_{n=1}^{\infty} a_n x^n \frac{dx}{x} \text{ is in } d(R[[x]]) \iff \frac{a_n}{n} \in R \text{ for all } n.$$

Thus,

$$\sum_{n=1}^{\infty} a_n x^n \frac{dx}{x} \in d(R[[x]]\otimes K) \iff \operatorname{ord}_p a_n \geq \operatorname{ord}_p n + \text{constant for all } n.$$

Evaluating F_0^* in (6.6) gives

$$e^{-\pi x^{pd}}\, x^{pa}\, p \frac{dx}{x} \equiv \lambda\, e^{-\pi x^d}\, x^{a'} \frac{dx}{x} \mod d(R[[x]]\otimes K). \tag{6.7}$$

Equating powers of x in (6.7) gives

$$\frac{(-\pi)^n}{n!}\, p \equiv \lambda \frac{(-\pi)^m}{m!} \mod p^{\operatorname{ord}_p(md+a') + \text{const}}, \tag{6.8}$$

where m is chosen so that $md + a' = p(nd + a)$, i.e., $m = \frac{pa - a'}{d} + pn$.

We now choose a sequence of n's for which $\operatorname{ord}_p(nd+a)$ (and hence $\operatorname{ord}_p(md+a')$) approaches infinity. Namely, we can let n

approach $-a/d$ by taking the p-adic expansion of $-a/d$; but it is more convenient to expand $-a/d$ in powers of $q = p^f$. Letting $b = \frac{q-1}{d}a$, so that $0 < b < q-1$, we have

$$-\frac{a}{d} = \frac{b}{1-q} = b + bq + bq^2 + \dots .$$

Let $n = b + bq + \dots + bq^{j-1} = -\frac{1-q^j}{d}a$. Then $\mathrm{ord}_p(nd+a) \geq fj$. Further note that $\mathrm{ord}_p\left(\frac{n!}{\pi^n}\right) = -\frac{S_n}{p-1} = -\frac{jS_b}{p-1}$. Similarly, $\mathrm{ord}_p\left(\frac{m!}{\pi^m}\right) = -\frac{jS_b}{p-1} + \text{const}$. Thus, multiplying (6.8) through by $m!(-\pi)^{-m}$ and carefully taking note of ord_p and the effect on the congruence, we obtain

$$\lambda \equiv p\,(-\pi)^{n-m}\frac{m!}{n!} \mod p^{fj - jS_b/(p-1) + \text{const}}$$

But $j\left(f - \frac{S_b}{p-1}\right) \longrightarrow \infty$ as $j \longrightarrow \infty$ (this is because $b < q-1$ and so has at least one digit less than $p-1$; thus, $S_b < f(p-1)$). Hence,

$$\lambda = p \lim_{j\to\infty} (-\pi)^{n-m}\frac{m!}{n!},$$

where $n = \frac{a}{d}(q^j-1)$, $m = pn + \frac{pa-a'}{d}$. Note that $\prod_{i\leq m,\ p|i} i = p\cdot 2p\cdots np = p^n n!$.

Thus, by the definition of Γ_p,

$$\Gamma_p(m+1) = (-1)^{m+1}\prod_{i\leq m,\ p\nmid i} i = (-1)^{m+1}\frac{m!}{n!\ p^n}.$$

Hence,

$$\lambda = p \lim_{j\to\infty} (-\pi)^{n-m}(-1)^{m+1}p^n\,\Gamma_p(m+1).$$

Now

$$(-\pi)^{n-m}(-1)^{m+1}p^n = -\pi^{n-m}(-p)^n = -\pi^{pn-m},$$

since $\pi^{p-1} = -p$. Since $pn - m = -\frac{pa-a'}{d}$; and $n \longrightarrow -\frac{a}{d}$ and $m \longrightarrow -\frac{a'}{d}$ as $j \longrightarrow \infty$, we conclude:

$$\lambda = -p\,\pi^{-(pa-a')/d} \lim_{m \to -a'/d} \Gamma_p(m+1),$$

and the proposition is proved.

Remarks. 1. For simplicity, suppose $p \equiv 1 \pmod{d}$. Then the theorem reads:

$$g(\tilde{\chi}_a, \psi_\pi) = -p\,\pi^{-a(p-1)/d}\,\Gamma_p(1-\tfrac{a}{d}). \tag{6.9}$$

Suppose $1 \le r,\ s,\ r+s < d$, and let $\overline{\chi}_a$ denote $\tilde{\chi}_{d-a}$. By property (3) of Gauss and Jacobi sums at the beginning of this chapter, (6.9) gives us:

$$J(\overline{\chi}_r, \overline{\chi}_s) = \frac{\Gamma_p\left(\frac{r}{d}\right)\,\Gamma_p\left(\frac{s}{d}\right)}{\Gamma_p\left(\frac{r+s}{d}\right)},$$

which looks remarkably similar to the beta function value in (2.2) for the classical periods of the differential $\omega_{r,s}$!

2. The above theorem also gives an analogy between the Chowla-Selberg formula for the periods of an elliptic curve with complex multiplication and a p-adic expression for the roots of the zeta function of the elliptic curve; see [37,34].

We have thereby shown Gauss sums to be p-adic analogs of special values of the gamma function. In the next section, we show how Stickleberger's theorem on the ideal decomposition of Gauss sums is an immediate corollary. In the next chapter we shall see a subtler application: the proof that the p-adic Dirichlet L-function $L_p(s,\chi)$ has at most a simple zero at $s = 0$.

7. Stickleberger's theorem

Stickleberger's theorem gives the ideal decomposition of Gauss sums $g(\chi,\psi)$ in $\overline{\mathbb{Q}}$. Let K denote the field $\mathbb{Q}(\sqrt[d]{1})$. Let P be any fixed prime ideal of K lying over p. Let q be the number of elements in the residue field O/P of P; thus, $q = p^f$ is the least power of p such that $d \mid q-1$. We identify $\mathbb{F}_q$ with O/P, and let

$$\chi\colon \mathbb{F}_q^* = (O/P)^* \longrightarrow \mu_d \subset K$$

be a multiplicative character. Let a be the integer, $0 \leq a < d$, determined by $\chi(x) \bmod P = x^{a(q-1)/d}$ for $x \in O/P$. Thus, if we use P to imbed K in Ω_p, and consider χ to take values in Ω_p, then χ is the $a\frac{q-1}{d}$-th power of the Teichmüller character, i.e., $\chi = \tilde{\chi}_a$ in the notation of §§3-6.

The Gauss sum $g(\chi,\psi)$ is obviously an algebraic integer in $K(\sqrt[p]{1})$. By checking the action of $\mathrm{Gal}(K(\sqrt[p]{1})/K)$ on

$$g(\chi,\psi) = -\sum_{x\in\mathbb{F}_q^*} \chi(x)\ \psi(x),$$

we see that $g(\chi,\psi)^d$ lies in K and is independent of the additive character ψ. By property (1) in §1, $g(\chi,\psi)$ divides $q = p^f$; hence, the ideal $\left(g(\chi,\psi)^d\right)$ in K must be a product of powers of prime ideals of K which divide p. Stickleberger's theorem gives these powers.

We can consider $g(\chi,\psi)^d$ p-adically if we choose an imbedding $\iota\colon K \hookrightarrow \mathbb{Q}_p(\sqrt[d]{1})$. As explained before (§II.2), for our fixed prime ideal P of K dividing p, we obtain such an imbedding $\iota = \iota_P$ by taking the completion of K in the P-adic topology; this imbedding ι allows us to identify χ (strictly speaking,

$\iota\circ\chi)$ with $\omega^{\frac{a}{d}(q-1)}$. The power of P dividing the ideal $\left(g(\chi,\psi)^d\right)$ in K is simply $\mathrm{ord}_p\left(\iota_p(g(\chi,\psi)^d)\right)$.

Recall that $\mathrm{Gal}(K/Q) \simeq (Z/dZ)^*$, where $\sigma_j: \xi \longmapsto \xi^j$ for $j \in (Z/dZ)^*$, $\xi^d = 1$. $\mathrm{Gal}(K/Q)$ permutes the prime ideals of K dividing p, and we let $P_j = \sigma_j P$. Of course, $P_j = P_{j'}$ if j/j' is in the subgroup of powers of p in $(Z/dZ)^*$ (the "decomposition group" of P).

If $a/b \in Q$ with $\mathrm{g.c.d.}(b,d) = 1$, let $\langle a/b\rangle_d$ denote the least positive residue of a/b mod d, i.e., the least positive k such that $kb \equiv a \pmod d$. Let

$$\left(g(\chi,\psi)^d\right) = \prod_{j \in (Z/dZ)^*/\{p^k\}} P_j^{\alpha_j}$$

be the ideal decomposition of $g(\chi,\psi)^d$.

Stickleberger's theorem. $\alpha_j = \sum_{k=0}^{f-1} \langle -a/jp^k\rangle_d$, *i.e.*,

$$\left(g(\chi,\psi)^d\right) = P^{\left(\sum_{j\in(Z/dZ)^*} \langle -a/j\rangle_d \sigma_j\right)},$$

where we write P^{σ_j} *for* $\sigma_j P = P_j$.

Proof. α_j = power of P_j dividing $g(\chi,\psi)^d$

= power of P dividing $\sigma_j^{-1} g(\chi,\psi)^d$

= power of P dividing $g(\chi^{(j^{-1})},\psi)^d$,

where $\chi^{(j^{-1})} = \sigma_j^{-1}\circ\chi$ is the character $\chi^{\langle 1/j\rangle_d} = \omega^{\langle a/j\rangle_d \frac{q-1}{d}}$. But according to the theorem in §6,

$$\mathrm{ord}_p(\iota_p(g(\chi^{(j^{-1})},\psi)^d)) = \sum_{k=0}^{f-1}(d - \langle p^k a/j\rangle_d) = \sum_{j=0}^{f-1} \langle -a/jp^k\rangle_d. \quad \text{Q.E.D.}$$

IV. p-ADIC REGULATORS

1. Regulators and L-functions

If K is a number field with r_1 real imbeddings and $2r_2$ complex imbeddings, a classical theorem of Dirichlet [13] asserts that the multiplicative group E of units of K is the direct product of the (finite) group W of roots of 1 in K and a free abelian group of rank r_1+r_2-1, i.e., there exist units $e_1,\ldots,$ $e_{r_1+r_2-1}$ such that every unit can be written uniquely in the form

$$\eta\, e_1^{m_1} e_2^{m_2} \cdots e_{r_1+r_2-1}^{m_{r_1+r_2-1}}, \quad m_i \in Z, \quad \eta \text{ a root of } 1.$$

If $\phi_1,\ldots,\phi_{r_1}$ denote the real imbeddings and $\phi_{r_1+1},\ldots,\phi_{r_1+r_2}$ denote the complex imbeddings (one chosen from each complex conjugate pair), then the map

$$\Big(\log|\phi_1(\ )|,\ldots,\ \log|\phi_{r_1}(\ )|,$$
$$2\log|\phi_{r_1+1}(\ )|,\ldots,\ 2\log|\phi_{r_1+r_2}(\ )|\Big):\ K \longrightarrow R^{r_1+r_2}$$

takes the group E/W of units modulo roots of 1 isomorphically to a lattice in the hyperplane $x_1+x_2+\cdots+x_{r_1+r_2}=0$ in r_1+r_2-dimensional real space. The volume

$$\det(n_i \log|\phi_i(e_j)|), \quad 1 \le i,j \le r_1+r_2-1,$$

where $n_i=1$ for $i \le r_1$, $n_i=2$ for $i > r_1$, of a fundamental parallelotope (actually of its projection on the $x_{r_1+r_2}$-hyperplane) is called the (classical) regulator of K. It depends only on K, not on the choice of "fundamental units" e_j or the ordering of the ϕ_i, and it is always nonzero.

In this chapter we discuss two very different types of p-adic regulators. The first type, due to Leopoldt, takes the same units e_j as in the classical case, replaces the imbeddings $\phi_i: K \hookrightarrow C$ by imbeddings ϕ_i of K into the algebraically closed field Ω_p, and replaces $\log$ by $\ln_p$. Because there is no natural way of eliminating r_2 of the p-adic ϕ_i the way we eliminated one ϕ_i from each complex conjugate pair in the complex case, Leopoldt further assumes that $r_2 = 0$, i.e., K is totally real.

The second, more recently developed type of p-adic regulator is due to B. H. Gross ([35], see also Greenberg [31]). It applies to number fields K which are totally complex, i.e., $r_1 = 0$; more specifically, to so-called CM fields, which are quadratic imaginary extensions of a totally real field K^+, i.e., $K = K^+(\sqrt{-\alpha})$ for some α which is positive under all imbeddings $K^+ \hookrightarrow R$. Gross works not with ordinary units, but rather with "p-units". A p-unit is an algebraic number which has absolute value 1 at all places (including archimedean valuations) except for those over p. In other words, all of its conjugates must have complex absolute value 1, and its ideal decomposition only involves primes dividing p. A key example of a p-unit is $g/\overline{g}$, where g is a Gauss sum for a finite field of $q = p^f$ elements. The multiplicative group of p-units, if tensored with Q and written additively, is isomorphic to the vector space of divisors $\sum_{P|p} Q\,(P-\overline{P})$. Among the imbeddings $\phi_i: K \hookrightarrow \Omega_p$, Gross chooses one from each coset modulo $\pm$ (the decomposition group of p), so that each ϕ_i gives a different permutation of the divisors $P - \overline{P}$. For a more precise statement, see below. Gross then takes the determinant of a $\frac{g}{2} \times \frac{g}{2}$ matrix, where g is the number of primes P over p. Gross's p-adic regulator is very different from Leopoldt's. In Gross's case the set of units considered and even the size of the matrix vary completely from one p to another.

The basic way in which regulators occur "in nature" is in the expansion at 1 or 0 of ζ- and L-functions. First, in the classical case, let

$$\zeta_K(s) = \sum \frac{1}{(NA)^s}$$

be the Dedekind zeta function of the number field K. Here the sum is over all non-zero integral ideals A of K, and N is the norm. The series converges for $\mathrm{Re}\, s > 1$ and can be analytically continued to a function which is holomorphic on the complex plane except for a simple pole at $s = 1$. The residue at $s = 1$ equals

$$\frac{2^{r_1}(2\pi)^{r_2} h R}{w \sqrt{|D|}},$$

where r_1 and r_2 are, as above, the number of real imbeddings and pairs of complex imbeddings; w is the number of roots of 1 in K, D is the discriminant of K, h is its class number, and R is its regulator. The subtlest and most elusive term in this formula is the regulator.

The Leopoldt p-adic regulator occurs in a similar way. Let K be a totally real field, i.e., $r_2 = 0$. Serre [85] has shown how to associate to K a p-adic zeta function $\zeta_{K,p}(s)$ which is defined and holomorphic on the closed unit disc in Ω_p (actually, on a slightly larger disc) except for a possible pole at $s = 1$. Conjecturally, the pole at $s = 1$ is a simple pole with residue given by

$$\frac{2^{r_1} h R_{p,\mathrm{Leopoldt}}}{w\sqrt{D}} E .$$

Here all of the terms have been defined above except for E, which is a product of Euler factors. (The phenomenon of "throwing out the p-Euler factor" can be expected to occur in all p-adic versions cf classical formulas for ζ- and L-functions.) In the simplest case, when $K = Q$, we have $r_1 = 1$, $h = 1$, $w = 2$, $D = 1$, $R_{p,\mathrm{Leopoldt}} = 1$, and $E = 1 - \frac{1}{p}$.

Note that there's an ambiguity of sign in $\sqrt{D}$. We will see that also $R_{p,\mathrm{Leopoldt}}$ is only defined up to a sign. In the classical case one can normalize by taking the absolute value of the determinant in the definition of R and the positive square root of $|D|$. It is harder to fix the sign in the p-adic case.

It is also conjectured that always $R_{p,\text{Leopoldt}} \neq 0$, i.e., there really is a pole at $s = 1$. Both the residue formula and the non-vanishing of the regulator have been proved in the case when K is an abelian extension of Q (the "abelian over Q" case). In that case ζ_K is a product of Dirichlet L-series, and the necessary facts were essentially worked out by Leopoldt [64] (see also [61]). We shall prove the non-vanishing of $R_{p,\text{Leopoldt}}$ below in the abelian over Q case. The proof uses a theorem from transcendence theory, which will be stated without proof.

But very little is known about $R_{p,\text{Leopoldt}}$ in the non-abelian case. A partial result supporting the residue formula was obtained by Serre [86], who proved that for any totally real field K, if $\zeta_{K,p}$ has a pole at 1, then $R_{p,\text{Leopoldt}} \neq 0$.

It should also be mentioned that the "Leopoldt conjecture" (non-vanishing of $R_{p,\text{Leopoldt}}$) and the expected relationship between the p-adic regulator and the residue at 1 has been generalized by Serre to p-adic "Artin L-functions" associated to representations of the Galois group of $\bar{k}/k$ (k totally real).

Gross's p-adic regulator, we shall see, is connected to the behavior near $s = 0$ of p-adic Artin L-functions. These are p-adic L-functions $L_p(s,\rho)$ which p-adically interpolate values of the Artin L-series associated with a representation ρ of the Galois group of a CM field K over a totally real ground field k. The order of zero m_ρ of $L_p(s,\rho)$ at $s = 0$ has been conjectured for some time (see [29]). Gross further conjectures that the leading term in the Taylor series at 0 of $L_p(s,\rho)$ is

$$R_{p,\text{Gross}}(\rho)\ A(\rho)\ s^{m_\rho},$$

where $R_{p,\text{Gross}}(\rho)$ is Gross's p-adic regulator and $A(\rho)$ is an explicitly given algebraic number, which turns out to be a product of certain Euler type factors and an algebraic number which is independent of p. (For a more detailed account, see below.) Note the analogy with the Leopoldt residue formula discussed above, in which the leading coefficient of the Laurent expansion (at $s=1$) is the product of a (p-adic transcendental) regulator, an Euler term, and an algebraic number independent of p. In the classical case, as we shall see below, the functional equation for L-functions gives a direct relationship between the expansion at $s=1$ and the expansion at $s=0$. But in the p-adic case there is no functional equation, and no one has yet been able to explain the analogy between the Leopoldt and the Gross formulas, in the sense of providing a direct link between the two types of p-adic regulators.

Gross's conjectured formula was motivated by: Ferrero-Greenberg's proof [29] that p-adic Dirichlet L-series have at most a simple zero at 0; and a conjecture of Stark and Tate concerning the leading coefficient at 0 of classical Artin L-series. Gross's conjecture is known to be true when K is an abelian extension of Q. In the abelian over Q case it reduces to the case when ρ is a one-dimensional character, $m_\rho = 1$, and the conjecture asserts that

$$L_p'(s,\rho) = R_{p,\mathrm{Gross}}(\rho)\ A(\rho) \neq 0 \quad \text{at} \quad s=0.$$

We shall give Gross's variant of Ferrero-Greenberg's original proof of this fact.

Gross developed his conjecture as a p-adic analog of a conjecture of Stark [90] and Tate [93]. Instead of giving the Stark-Tate conjecture in the general setting, I'll illustrate the idea by showing how it interprets the classical formula (see §II.5)

$$L(1,\chi) = -\frac{g_\chi}{d} \sum_{0<a<d} \bar{\chi}(a) \log(1-\zeta^{-a}), \tag{1.1}$$

where χ is a nontrivial Dirichlet character of conductor d, ζ is a primitive d-th root of 1, and $g_\chi = \Sigma\, \chi(a)\, \zeta^a$.

For simplicity, we take the case when χ is a nontrivial <u>even</u> character and $d = p^N$ is a power of an odd prime p.

Let $K = Q(\zeta)$, where ζ is a primitive d-th root of 1, and let $K^+ = Q(\zeta + \zeta^{-1})$ be its maximal totally real subfield. Then $Gal(K/Q) \approx (Z/dZ)^*$, $\sigma_a(\zeta) = \zeta^a$, and $Gal(K^+/Q) \approx (Z/dZ)^*/\{\pm 1\}$. Let G denote $(Z/dZ)^*/\{\pm 1\}$, so that summation over $a \in G$ means taking representatives from half of the residue classes in $(Z/dZ)^*$.

Let g be a generator of the cyclic group $(Z/dZ)^*$ (recall that d is an odd prime power), so that $\sigma_g \in Gal(K^+/Q)$ generates $Gal(K^+/Q)$. Let

$$\varepsilon = (\zeta - \zeta^{-1})^{\sigma_g - 1} = \frac{\zeta^g - \zeta^{-g}}{\zeta - \zeta^{-1}}.$$

Then it can be shown ([61], p. 85) that ε is a Minkowski unit in K^+ (also in K), i.e., $\{\sigma_a \varepsilon\}_{a \in G}$ generate a subgroup of finite index in the group E of units of K^+. Let $E_{cyc} \subset E$ denote this group of "cyclotomic units". (Equivalently, the $\sigma_a \varepsilon$ are multiplicatively independent except for the single relation $\Pi_{a \in G} \sigma_a \varepsilon = N\varepsilon = 1$. The situation is a little messier when d is not a prime power.)

Let $C[G]$ be the group-ring over the complex numbers of $G = (Z/dZ)^*/\{\pm 1\}$. Let I be the ideal generated by the element $\Sigma_{\sigma \in G} \sigma$. Then it is easy to see that E_{cyc} is a free rank-one $Z[G]/I$-module with generator ε, where we define

$$\varepsilon^{\Sigma a_\sigma \sigma} = \prod (\sigma\varepsilon)^{a_\sigma}, \quad \sum a_\sigma \sigma \in Z[G]/I;$$

and $E_{cyc} \otimes C$ is a free rank-one $C[G]/I$-module.

Let X denote $C[G]/I$. Let $LOG: X \longrightarrow X$ be the map defined on a basis element by

$$LOG(\sigma_a) = \sum_{b \in G} \log|\sigma_b \sigma_a \varepsilon| \, \sigma_b^{-1}.$$

The determinant of LOG is clearly the regulator of K^+ (times the index $[E:E_{cyc}]$). We can write the map LOG explicitly as

$$LOG(\sigma_a) = \sum_{b \in G} \log\left|\frac{\zeta^{abg} - \zeta^{-abg}}{\zeta^{ab} - \zeta^{-ab}}\right| \sigma_b^{-1}.$$

It is easy to check that this is a well-defined map from X to X.

An irreducible representation of $C[G]/I$ is the same as a non-trivial even character mod d. Let χ be such a character. Define the χ-regulator R_χ to be the determinant of the map induced by LOG on $V_\chi \bigotimes_{C[G]} X$, where G acts on the 1-dimensional space V_χ by χ. Here $V_\chi \bigotimes_{C[G]} X$ can simply be identified with the χ-eigen space of X, i.e., the 1-dimensional subspace spanned by $\Sigma\,\bar{\chi}(a)\,\sigma_a$. Then

$$\mathrm{LOG}\left(\Sigma\,\bar{\chi}(a)\,\sigma_a\right) = \sum_{a,b\in G} \bar{\chi}(a)\ \log|\sigma_b\sigma_a\varepsilon|\ \sigma_b^{-1}$$

$$= \sum_{b,c\in G} \chi(b)\,\bar{\chi}(c)\ \log|\sigma_c\varepsilon|\ \sigma_{b^{-1}} \qquad (c = ab)$$

$$= \left(\sum_{a\in G} \bar{\chi}(a)\ \log|\sigma_a\varepsilon|\right)\left(\sum_{b\in G} \bar{\chi}(b)\ \sigma_b\right).$$

Thus

$$R_\chi = \sum_{a\in G} \bar{\chi}(a)\ \log\left|\frac{\zeta^{ag} - \zeta^{-ag}}{\zeta^{a} - \zeta^{-a}}\right|$$

$$= \sum_{a\in G} \bar{\chi}(a)\ (\log|1-\zeta^{-2ag}| - \log|1-\zeta^{-2a}|)$$

$$= \sum_{a\in G} \chi(2g)\,\bar{\chi}(2ag)\ \log|1-\zeta^{-2ag}| - \sum_{a\in G}\chi(2)\,\bar{\chi}(2a)\ \log|1-\zeta^{-2a}|$$

$$= \chi(2)\ (\chi(g)-1) \sum_{a\in G} \bar{\chi}(a)\ \log|1-\zeta^{-a}|$$

$$= \frac{\chi(2)}{2}\ (\chi(g)-1) \sum_{a=1}^{d} \bar{\chi}(a)\ \log(1-\zeta^{-a})$$

(because χ is even, and $\log(1-\zeta^{-a})(1-\zeta^{a}) = \log|1-\zeta^{-a}||1-\zeta^{a}|$). Comparing with (1.1), we see that the only difference between R_χ and $L(1,\chi)$ is an algebraic factor; that is, R_χ is the transcendental part of $L(1,\chi)$. It is this fact which Stark and Tate generalize in their conjecture.

We get a companion fact about the behavior of $L(s,\chi)$ near $s=0$ if we use the functional equation for $L(s,\chi)$, which relates $L(s,\chi)$ to $L(1-s,\bar{\chi})$, and hence relates behavior near $s=1$ to behavior near $s = 0$. Suppose χ is a nontrivial even character. Then the functional equation is (see, e.g., [41], p. 5):

$$L(s,\chi) = \frac{g_\chi}{2}\left(\frac{2\pi}{d}\right)^s \frac{L(1-s,\bar{\chi})}{\Gamma(s)\cos(s/2)} .$$

If we let $s \longrightarrow 0$ and write $\Gamma(s) = \Gamma(s+1)/s$, we find that $L(0,\chi) = 0$ (which we knew, since $B_{1,\chi} = 0$ for χ even and nontrivial), and the Taylor expansion at $s = 0$ starts out

$$L(s,\chi) = s\,\frac{g_\chi}{2}\,L(1,\bar{\chi}) \quad + \quad \text{higher terms.}$$

Hence, the transcendental part of the first nonzero Taylor coefficient is the <u>same</u> as the transcendental part of $L(1,\bar{\chi})$, i.e., it is $R_{\bar{\chi}}$. Note that the non-vanishing of $R_{\bar{\chi}}$ implies non-vanishing of $L(1,\bar{\chi})$, and at 0 it implies that the zero of $L(s,\chi)$ is simple, i.e., $L'(0,\chi) \neq 0$.

More generally, the Stark-Tate conjecture can be stated equivalently in terms of the behavior near either 1 or 0, thanks to the functional equation.

In the p-adic case, there is no known (or expected) functional equation, and so there are two <u>completely different</u> p-adic analogs of the Stark-Tate conjecture, one at $s = 1$ (due to Leopoldt and Serre), and one at $s = 0$ (due to Gross).

2. Leopoldt's p-adic regulator

Let K be a totally real number field, $n = [K:Q]$. By Dirichlet's unit theorem, the group E of units of K is the product of the roots of 1 in K and a free abelian group of rank $n-1$. Let $e_1,\ldots,e_{n-1}$ be generators of this free abelian group. Let $\phi_1,\ldots \phi_n\colon K \hookrightarrow \Omega_p$ be all of the n possible imbeddings of K into the algebraically closed field Ω_p. The Leopoldt (p-adic) regulator of K is defined as the determinant of the $(n-1)\times(n-1)$ matrix

$$\{\ln_p \phi_i(e_j)\}_{1\le i,\,j\le n-1}. \tag{2.1}$$

Lemma. <u>The Leopoldt regulator</u> $R = R_{p,\text{Leopoldt}}(K)$ <u>is independent up to</u> ± 1 <u>of the choice of basis</u> $\{e_j\}$ <u>and the ordering of</u>

the ϕ_i.

Proof. Any other basis $e' = \{e'_j\}$ can be written in the form $e' = e^M$, where M is an $(n-1)\times(n-1)$ matrix $\{m_{kj}\}$ with $m_{kj} \in \mathbb{Z}$ and $\det M = \pm 1$. The notation $e' = e^M$ here means that $e'_j = \prod_k e_k^{m_{kj}}$. Then also $\phi_i(e'_j) = \prod_k \phi_i\left(e_k^{m_{kj}}\right)$, and $\ln_p \phi_i e' = \left(\ln_p \phi_i e\right) M$, i.e., $\ln_p \phi_i(e'_j) = \sum_k m_{kj} \ln_p \phi_i(e_k)$. This means that replacing e by e' in (2.1) amounts to multiplying the matrix (2.1) by M on the right. Since $\det M = \pm 1$, the independence of choice of basis e is clear.

Rearranging the $\phi_1, \ldots, \phi_{n-1}$ clearly changes the determinant at most by a sign. It remains to consider what happens if some ϕ_k, $1 \le k \le n-1$, changes places with ϕ_n. Since $\prod_{i=1}^{n} \phi_i(e_j) = 1$ for any unit e_j, we have $\sum_{i=1}^{n} \ln_p \phi_i(e_j) = 0$, and so adding all the other rows to the k-th row gives $\sum_{i=1}^{n-1} \ln_p \phi_i(e_j) = -\ln_p \phi_n(e_j)$ in the k-th row; hence interchanging ϕ_k and ϕ_n only changes the determinant by a minus sign, and the lemma is proved.

The Leopoldt conjecture. $R_{p,\mathrm{Leopoldt}}(K) \neq 0$ *for any totally real number field* K.

In the simplest case, when $n = 2$, i.e., K is real quadratic, the non-vanishing of $R_{p,\mathrm{Leopoldt}}(K)$ simply says that $\ln_p$ of a fundamental unit e is nonzero. Since the kernel of the $\ln_p$ map consists of powers of p times roots of unity, while ord_p of any unit is 0 and e is not a root of 1, it immediately follows that $\ln_p e \neq 0$.

More generally, we shall prove Leopoldt's conjecture for all abelian extensions K of $\mathbb{Q}$. The proof relies upon the p-adic version of the following deep theorem of transcendence theory.

Baker's theorem [9]. *If* $0 \neq \alpha_i \in \overline{\mathbb{Q}} \subset \mathbb{C}$ *and* $\{\log \alpha_i\} \subset \mathbb{C}$ *are linearly independent over* $\mathbb{Q}$, *then* $\{\log \alpha_i\}$ *are linearly* independent over $\overline{Q}$.

But before showing how the Leopoldt conjecture for K/Q abelian can be derived from the p-adic Baker theorem, we first give another interpretation of this conjecture, in terms of which one can state a natural more general conjecture.

Let K be a number field with ring of integers 0 and group of units E. Let $[K:Q] = n = r_1 + 2r_2$, where r_1 $(2r_2)$ is the number of real (complex) imbeddings. Let P_i, $i = 1,\dots,g$, be the primes of K dividing p, and let $0_{P_i} \subset K_{P_i}$ be the P_i-adic completion of $0 \subset K$. Let N_i denote the norm from K_{P_i} to Q_p.

Let

$$A = \prod_{i=1}^{g} 0^*_{P_i}, \qquad A_0 = \{(x_1,\dots,x_g) \in A \mid \Pi N_i(x_i) = 1\}.$$

Let $E_0 \subset E$ be the subgroup of units of norm $+1$; E_0 has index 2 in E if 0 has units with norm -1, otherwise $E_0 = E$. Then $E_0 \subset 0 \hookrightarrow 0_{P_i}$ imbeds in A_0.

Let $\overline{E}_0$ be the closure of E_0 in A_0. To get a concrete idea of what $\overline{E}_0$ looks like, let $e_1,\dots,e_{r_1+r_2-1} \in E$ be a set of fundamental units of norm $+1$, i.e., they generate E_0 modulo roots of 1. Thus, $E_0 = \{\eta \Pi e_j^{\alpha_j} \mid \alpha_j \in Z,\ \eta \text{ a root of } 1\}$. Now let N be an integer such that $e_j^N \equiv 1 \pmod{P_i}$ for all i and j. For example, N can be chosen to be $\prod_{i=1}^{g}(q_i - 1)$, where $q_i = p^{f_i}$ is the number of elements in the residue field $0/P_i$. Let $e'_j = e_j^N$. Then e'_j can be raised to p-adic powers $\alpha_j \in Z_p$ in A_0, because its image in each $0^*_{P_i}$ is close to 1. It is easy to see that $\overline{E}_0 \subset A_0$ is precisely the set of elements of the form $\eta \Pi e_j^{\beta_j} e'^{\alpha_j}_j$, where η is a root of 1, $0 \le \beta_j < N$, and $\alpha_j \in Z_p$.

Proposition. <u>For</u> K <u>totally real, Leopoldt's conjecture is</u>

equivalent to the assertion that $\overline{E}_0$ is a subgroup of finite index in A_0.

Thus, if K has r_2 pairs of complex imbeddings, a natural generalization of Leopoldt's conjecture is: $A_0/\overline{E}_0$ is isomorphic to (finite group) $\times Z_p^{r_2}$.

Proof of proposition. First note that the fundamental units e_j can be raised to some power $e_j^{Np^M}$ such that the image of $e_j^{Np^M}-1$ in O_{P_i} has p-adic absolute value less than $p^{-1/(p-1)}$ for all i, j. For example, if all of the P_i are unramified (and $p > 2$), then we need only take e_j^N, where N is chosen as above so that $e_j^N \equiv 1 \pmod{P_i}$ for all i, j. Let $e_j' = e_j^{Np^M}$, let $E' = \{\Pi e_j'^{\alpha_j} | \alpha_j \in Z\} \subset E_0$, $\overline{E}' = \{\Pi e_j'^{\alpha_j} | \alpha_j \in Z_p\} \subset \overline{E}_0$. Then E' has finite index in E_0, and $\overline{E}'$ has finite index in $\overline{E}_0$.

Note that, if we replace e_j by e_j' in the definition of $R = R_{p,\text{Leopoldt}}$, obtaining a new determinant R', the effect is to multiply the regulator by a nonzero constant (in fact, by $(Np^M)^{n-1}$ or $2(Np^M)^{n-1}$, since each entry in the $(n-1)\times(n-1)$ matrix is multiplied by Np^M, and we also have to throw in a 2 if $E_0 \neq E$). Thus, the proposition is equivalent to: $R' = 0$ if and only if $[A_0 : \overline{E}'] < \infty$.

Let $n_i = [K_{P_i} : Q_p]$ be the local degree, and let σ_{it}, $t = 1, \ldots, n_i$, be the imbeddings $K_{P_i} \hookrightarrow \Omega_p$. Let $K_{P_i,t} = \sigma_{it}K_{P_i}$, $O_{P_i,t} = \sigma_{it}O_{P_i}$. Let $B = \Omega_p^n$, and let $B_0 = \{(y_1,\ldots,y_n) \in B | \Sigma y_i = 0\}$. Define LOG: $A \twoheadrightarrow B$ by

$$\text{LOG}(x_1,\ldots,x_g) = (\ln_p\sigma_{11}(x_1),\ldots,\ln_p\sigma_{1,n_1}(x_1),\ldots,\ln_p\sigma_{g1}(x_g),\ldots,\ln_p\sigma_{g,n_g}(x_g)).$$

Let $O'_{P_i} = \{x \in O^*_{P_i} \mid |x-1|_p < p^{-1/(p-1)}\}$, which is a subgroup of finite index in $O^*_{P_i}$. (Since $|\sigma_{it} x - 1|_p$ is independent of the imbedding, we denote it $|x-1|_p$.) Let $1 + \pi_i \in O'_{P_i}$ be an element with $|\pi_i|_p$ maximal, and denote $\pi_{it} = \sigma_{it}(\pi_i)$. We claim that $\ln_p$ maps $O'_{P_i,t} = \sigma_{it} O'_{P_i}$ isomorphically to $\pi_{it} O_{P_i,t}$. To see this, first check that for any $\alpha \in O_{P_i,t}$ the series

$$(1 + \pi_{it})^\alpha = \sum \binom{\alpha}{j} \pi_{it}^j$$

converges to an element of $O'_{P_i,t}$ whose p-adic distance from 1 is $|\alpha\pi_{it}|_p$. Since $\ln_p$ and the exponential function give mutually inverse isomorphisms between the open disc of radius $p^{-1/(p-1)}$ around 1 and the open disc of radius $p^{-1/(p-1)}$ around 0, it follows that $O'_{P_i} = (1+\pi_i)^{O_{P_i}} \xrightarrow[\approx]{\ln_p \circ \sigma_t} \pi_{it} O_{P_i,t}$.

Thus, $\prod_{i=1}^{g} O'_{P_i}$ is a subgroup of finite index in A which is taken isomorphically by LOG to the free rank-one $\bigoplus_{i=1}^{g} O_{P_i}$-module in B generated by $(\ldots \ln_p(\pi_{it}) \ldots)_{i=1,\ldots,g; t=1,\ldots,n_i}$. Since $\operatorname{rank}_{Z_p} \bigoplus O_{P_i} = n$, it follows that $\operatorname{rank}_{Z_p} \mathrm{LOG}(A) = n$, and $\operatorname{rank}_{Z_p} \mathrm{LOG}(A_0) = n-1$. Since LOG maps $\overline{E'} \subset \prod O'_{P_i}$ isomorphically to the Z_p-submodule of B_0 spanned by

$$\{(\ldots \ln_p \sigma_{it} e'_j \ldots)_{i,t}\}_{j=1,\ldots,n-1}, \tag{2.2}$$

it follows that

$$[A_0 : \overline{E'}] < \infty \iff [\mathrm{LOG}(A_0) : \mathrm{LOG}(\overline{E'})] < \infty$$

$$\iff \operatorname{rank}_{Z_p} \mathrm{LOG}(\overline{E'}) = n-1$$

$\iff$ the set of vectors (2.2) has rank $n-1$

$\iff R' \neq 0.$ Q.E.D.

The proposition just proved can be paraphrased roughly as

follows: Leopoldt's conjecture says that a system of fundamental units is independent not only over Z, but even over Z_p.

We now prove Leopoldt's conjecture in the abelian over Q case.

Theorem. Let K be a totally real abelian extension of Q. Then $R_{p,\text{Leopoldt}}(K) \neq 0$.

Proof. Fix an imbedding $\phi: K \hookrightarrow \Omega_p$. Let $G = \text{Gal}(K/Q)$, $n = [K:Q]$. Then the imbeddings $\phi_i: K \hookrightarrow \Omega_p$ are $\{\phi \circ \sigma\}_{\sigma \in G}$. Let σ_0 be any fixed element of G. Let $\{e_j\}$ be a basis of the units of K. Then

$$R = R_{p,\text{Leopoldt}}(K) = \text{Det}\,\{\ln_p \phi\sigma(e_j)\}_{j=1,\dots,n-1;\ \sigma \in G-\{\sigma_0\}}.$$

If $R = 0$, then the rows of this matrix are linearly dependent over Ω_p, i.e.,

$$\sum_{\sigma \in G-\{\sigma_0\}} a_\sigma^0 \ln_p \phi\sigma(e_j) = 0, \qquad j = 1,\dots,n-1,$$

for some $a_\sigma^0 \in \Omega_p$ not all zero. Since any unit e is a root of 1 times a product of the e_j, we have

$$\sum_{\sigma \in G} a_\sigma^0 \ln_p \phi\sigma(e) = 0, \qquad a_{\sigma_0}^0 = 0, \qquad \text{for all units } e. \tag{2.3}$$

Let $\Omega_p[G]$ be the group-ring over Ω_p of G. Define

$$I = \{\sum_{\sigma \in G} a_\sigma \sigma \in \Omega_p[G] \mid \sum_{\sigma \in G} a_\sigma \ln_p \phi\sigma(e) = 0 \text{ for all units } e\}.$$

Then I is an ideal of $\Omega_p[G]$, since it is clearly closed under addition, and for any $\tau \in G$, $\Sigma a_\sigma \sigma \in I \Longrightarrow \sum_\sigma a_\sigma \ln_p \phi\sigma(\tau e) = 0$ for all units $e \Longrightarrow \tau \Sigma a_\sigma \sigma \in I$. By (2.3), $\Sigma a_\sigma^0 \sigma$ is an element of I. Since $a_{\sigma_0}^0 = 0$, this element is not a multiple of $\Sigma \sigma$. Hence, we can find a nontrivial character $\chi: G \longrightarrow \Omega_p^*$ such that $\Sigma a_\sigma^0 \chi^{-1}(\sigma) \neq 0$. (This is because the function $f(\sigma) = a_\sigma^0$ on G can be expanded as a linear combination of characters of G: $f = \sum_\chi c_\chi \chi$, with $c_\chi = \frac{1}{n} \sum_\sigma a_\sigma^0 \chi^{-1}(\sigma)$; if $c_\chi = 0$ for all nontrivial χ, then f would be a multiple of the trivial character, and $\Sigma a^0_\sigma \sigma$ would be a multiple of $\Sigma\sigma$.)

So let χ be such that $\Sigma a^0_\sigma \chi^{-1}(\sigma) \neq 0$, and let

$$\sigma_\chi = \sum \chi(\sigma)\, \sigma . \tag{2.4}$$

Then, since I is an ideal, it contains

$$\sigma_\chi \sum a^0_\sigma \sigma = \sum_\sigma \sigma \sum_\tau a^0_\tau \chi(\tau^{-1}\sigma) = \left(\sum a^0_\sigma \chi^{-1}(\sigma)\right) \sigma_\chi .$$

Since the coefficient is nonzero, it follows that $\sigma_\chi \in I$. Note that also $\sigma_1 = \Sigma\sigma \in I$, because $\Pi\,\sigma(e) = 1$ for all units e. Thus, I contains $\sigma_\chi - \sigma_1 = \sum_{\sigma \neq id} (\chi(\sigma) - 1)\,\sigma$, i.e.,

$$\sum_{\sigma \neq id} (\chi(\sigma) - 1)\, \ln_p \phi\sigma(e) = 0 \quad \text{for all units } e. \tag{2.5}$$

We now use the p-adic version of Baker's theorem, which was proved by Brumer [15]. It is the same theorem, except that log is replaced by $\ln_p$, C is replaced by Ω_p, and we fix an imbedding of $\overline{Q}$ in Ω_p instead of C. That is,

p-adic Baker theorem. *If* $0 \neq \alpha_i \in \overline{Q} \subset \Omega_p$ *and* $\{\ln_p \alpha_i\} \subset \Omega_p$ *are linearly independent over* Q, *then they are linearly independent over* $\overline{Q}$.

Because of this theorem, we may conclude from (2.5) that for all e the set $\{\ln_p \phi\sigma(e)\}_{\sigma \in G-\{id\}}$ is linearly dependent over Q, i.e., over Z. Thus, for every e there are integers m_σ with $m_{id} = 0$ such that $\ln_p \phi\left(\prod_{\sigma \in G} \sigma(e)^{m_\sigma}\right) = 0$, i.e., $\phi\left(\Pi\sigma(e)^{m_\sigma}\right)$ is a power of p times a root of 1. Since ord_p of any unit is zero, $\Pi\,\sigma(e)^{m_\sigma}$ must be a root of 1. Thus, replacing m_σ by a multiple, we obtain: for each e there exist m_σ not all zero, but with $m_{id} = 0$, such that $\Pi\,\sigma(e)^{m_\sigma} = 1$.

But, by a theorem of Minkowski ([73], p. 90), there exists a unit e such that $\Pi\,\sigma(e)^{m_\sigma} = 1$, $m_{id} = 0$, implies that all of the $m_\sigma = 0$. That is, there exist units whose conjugates are multiplicatively independent except for the single relation $Ne = \Pi\sigma(e) = 1$. This contradiction proves the theorem.

The Leopoldt conjecture for all totally real fields would follow from the following conjecture in transcendence theory.

Conjecture (Schanuel). *If* $\alpha_1,\ldots,\alpha_r \in C$ *are linearly independent over* Q, *then*

$$\text{Tr.deg.}_Q\ Q(\alpha_1,\ldots,\alpha_r,\ e^{\alpha_1},\ldots,e^{\alpha_r}) \geq r.$$

The same holds if $\alpha_1,\ldots,\alpha_r \in \Omega_p$ *are in the disc of convergence of the* p-*adic exponential function and are linearly independent over* Q.

To see how Leopoldt's conjecture would follow from Schanuel's conjecture, we shall suppose that K is Galois (the general case can readily be reduced to the Galois case), in which case Minkowski's theorem cited above ensures the existence of a unit e which together with its conjugates $\sigma_i(e)$, $\sigma_i \in \mathrm{Gal}(K/Q)$, generates a subgroup of finite index in the unit group. Let $\phi_i = \phi\circ\sigma_i$ be the imbeddings $K \hookrightarrow \Omega_p$. In Schanuel's conjecture let $r = n-1$, $\alpha_i = \ln_p\phi_i(e)$, $i = 1,\ldots,n-1$. Replacing the full unit group by the subgroup generated by the $\phi_i(e)$ only changes the regulator by a nonzero constant multiple. If we set $e_j = \phi_j(e)$, we have $\phi_i(e_j) = \phi(\sigma_i\sigma_j(e))$. The regulator for the e_j is then the determinant of a matrix each of whose rows is a permutation of

$$\alpha_1 \quad \alpha_2 \quad \cdots \quad \alpha_{n-1} \quad (-\alpha_1-\alpha_2-\cdots-\alpha_{n-1})$$

with one entry omitted. Schanuel's conjecture says that $\alpha_1,\ldots,\alpha_{n-1}$ are algebraically independent. But vanishing of the regulator would give a nontrivial algebraic relation between the α's. (The easiest way to see the non-triviality of the polynomial in $\alpha_1,\ldots,\alpha_{n-1}$ is to note that if it were the zero polynomial, then the classical regulator, which is the same determinant with $\alpha_i = \log|\phi(\sigma_i(e))|$, $\phi\colon K \hookrightarrow C$, would also vanish, and it is well known that the classical regulator is nonzero.)

3. Gross's p-adic regulator

Let $K \subset C$ be a Galois extension of a totally real field k. Let τ be complex conjugation. Suppose we have an imbedding $\phi: K \hookrightarrow \Omega_p$ which extends a fixed imbedding $k \hookrightarrow \Omega_p$. Then any other such imbedding is of the form $\phi\circ\sigma$, $\sigma \in Gal(K/k)$. By a <u>complex conjugation</u> of $\phi(K) \subset \Omega_p$ we mean the image of τ under any of these imbeddings $\phi\circ\sigma$, i.e., any of the automorphisms $(\phi\circ\sigma)\circ\tau\circ(\phi\circ\sigma)^{-1} = \phi\circ(\sigma\tau\sigma^{-1})\circ\phi^{-1}$ of $\phi(K)$. If K/k is abelian, then of course there is only one complex conjugation $\phi\circ\tau\circ\phi^{-1}$ of $\phi(K)$.

Let k be totally real. Suppose we are given a representation

$$\rho: \; Gal(\bar{k}/k) \longrightarrow Aut(V),$$

where V is a finite dimensional vector space over Ω_p, which is trivial on $Gal(\bar{k}/K) \subset Gal(\bar{k}/k)$, i.e., ρ can be considered as a representation of the quotient $Gal(K/k)$. If ρ of any complex conjugation is the automorphism 1 (resp. -1), it is said to be an <u>even</u> (resp. <u>odd</u>) representation.

Using results of Deligne and Ribet [21], one can associate a p-adic L-function $L_p(s,\rho)$ to any even representation ρ. (If ρ is not even, the associated p-adic L-function is identically zero.) $L_p(s,\rho)$ is a meromorphic function from Z_p to Ω_p. It is conjectured to be holomorphic, except for a pole at $s=1$ when ρ contains the trivial representation. $L_p(s,\rho)$ is called the "p-adic Artin L-series associated to ρ."

Example. Let $k = Q$ and $K = Q(\zeta)$, where ζ is a primitive d-th root of 1. Let $\dim V = 1$, i.e., ρ is a one-dimensional character. Such characters ρ correspond to Dirichlet characters $\chi: (Z/dZ)^* \longrightarrow \Omega_p^*$ by the correspondence

$$\rho(\sigma_j)\, e = \chi(j)\, e, \quad \sigma_j \in Gal(K/Q) \approx (Z/dZ)^*,$$

where e is a basis of $V = \Omega_p e$ and $j \in (Z/dZ)^*$ is determined by σ as usual by $\sigma_j(\zeta) = \zeta^j$. Then ρ is even (resp. odd) if

$\chi(-1) = 1$ (resp. $\chi(-1) = -1$). In this case the p-adic Artin L-series associated to ρ is simply the p-adic Dirichlet L-series $L_p(s,\chi)$ which we studied in Chapter II.

In this example, recall that $L_p(s,\chi)$ p-adically interpolates the algebraic numbers

$$L_p(1-k,\chi) = (1 - \chi\omega^{-k}(p)\, p^{k-1})\, L(1-k, \chi\omega^{-k}),$$

where ω is the Teichmüller character. Gross's conjecture concerns the expansion near $s=0$ of the p-adic Artin L-series $L_p(s,\rho)$. In the present example note that when $1-k=0$ in the above formula, the p-adic L-function is related to the classical L-function for the character $\chi\omega^{-1}$. Specifically,

$$L_p(0,\chi) = (1-\chi\omega^{-1}(p))\, L(0,\chi\omega^{-1}) = (1-\chi\omega^{-1}(p))\,(-B_{1,\chi\omega^{-1}}) \quad (3.1)$$

If χ is even, then $\chi\omega^{-1}$ is odd. The Gross conjecture for the expansion of $L_p(s,\rho)$ near $s=0$, which in some sense is a vast generalization of (3.1), will thus involve expressions associated to the <u>odd</u> representation $\rho \otimes \omega^{-1}$.

It will take us a while to work up to the precise statement of Gross's conjecture. We first define the "p-units" of a number field K:

$$E = E^{(p)}(K) \underset{\text{def}}{=} \{e \in K \mid |e|_v = 1 \text{ for all valuations } v \nmid p\}.$$

This means that (1) in the factorization of the fractional ideal $(e) = \Pi P^{m_P}$ only $P|p$ occur; and (2) under all imbeddings $K \hookrightarrow \mathbb{C}$, e has complex absolute value 1. Note that the p-units are <u>not</u> contained in the ring of integers $\mathcal{O} \subset K$. Condition (2) means that the m for P must be negative the m for any complex conjugate prime ideal $\sigma\tau\sigma^{-1}(P)$. If all of the $m=0$, then it is well known that condition (2) implies that e is a root of 1.

In the above example, when $K = \mathbb{Q}(\zeta)$, an example of a p-unit is a ratio of Gauss sums of the form (see Chapter III)

$$g(\tilde{\chi}_a, \psi_\pi \circ \mathrm{Tr})^d / g(\tilde{\chi}_a^{-1}, \psi_\pi \circ \mathrm{Tr})^d.$$

We shall see that these p-units play a crucial role in the case when K is abelian over Q, which is the one case where Gross's conjecture is proved.

The basic case which is of interest is when K is a CM ("complex multiplication") field, i.e., a quadratic imaginary extension of a totally real field K^+. In that case there is only one complex conjugation, namely, the unique nontrivial element τ of $Gal(K/K^+)$, and we denote $\bar{\alpha} = \tau\alpha$ for $\alpha \in K$ and $\bar{P} = \tau P$ for a prime ideal P of K. Thus, if K is a CM field, (1) and (2) give

$$e \in E^{(p)}(K) \implies (e) = \prod_{P|p} (P/\bar{P})^{m_P}. \tag{3.2}$$

Writing (e) additively gives a homomorphism

$$E^{(p)}(K) \longrightarrow \bigoplus Z\,(P - \bar{P}),$$

where the sum is over primes P of K dividing p, one from each complex conjugate pair. The kernel of this homomorphism is the group of roots of 1 in K, and the image certainly contains $\bigoplus h\,Z\,(P - \bar{P})$, where h is the class number of K, because, if we write the principal ideal $P^h = (\alpha)$, we have

$$E^{(p)}(k) \ni \alpha/\bar{\alpha} \longmapsto h\,(P - \bar{P}).$$

Thus, if we tensor the Z-module $E = E^{(p)}(K)$ (i.e., the abelian group with respect to multiplication, which we write additively) with Q (thereby killing roots of 1), we obtain a Q-vector space

$$E \underset{Z}{\otimes} Q \approx \bigoplus Q\,(P - \bar{P}).$$

We shall want to adjust the above homomorphism $E \to \bigoplus Z(P - \bar{P})$ given by $e \longmapsto (e) = \Sigma\, m_P\,(P - \bar{P})$. Namely, at each P insert the residue degree $f_P = [O/P: F_p]$, where O is the ring of integers of K and F_p is the field of p elements. Also, insert a minus sign. Thus, let

$$\Psi(e) \underset{def}{=} -\sum m_P f_P\,(P - \bar{P}).$$

This map Ψ extends to an isomorphism $\Psi: E \otimes Q \xrightarrow{\sim} \bigoplus Q\,(P - \bar{P})$.

It is not hard to construct the inverse Φ of the "divisor map" Ψ. Let h be the class number of K, and write $P^h = (\alpha)$.

Let $e = \bar{\alpha}/\alpha \in E$. Then e is determined by P up to a root of unity, and so the element

$$\Phi(P - \bar{P}) = \frac{1}{hf_P} e \in E \otimes Q \tag{3.3}$$

is well-defined. Extend Φ by linearity to $\bigoplus Q(P - \bar{P})$. Clearly Φ and Ψ are inverse to one another. These maps allow us to think of $E \otimes Q$ as divisors, and to think of any additive function on E which kills roots of unity (for example, $\ln_p$) as a function on the divisors $\bigoplus Q(P - \bar{P})$.

We now define a function LOG (not the same function as the LOG in the preceding section) by letting $\phi: K \hookrightarrow \Omega_p$ run through all imbeddings, letting P_ϕ be the prime ideal dividing p which is defined by

$$P_\phi = \{x \in O \mid |\phi(x)|_p < 1\}, \tag{3.4}$$

and setting

$$\mathrm{LOG}(e) = \sum_\phi \ln_p \phi(e)\, P_\phi, \quad e \in E^{(p)}(K).$$

Combining terms with the same P_ϕ, we have

$$\mathrm{LOG}(e) = \sum_P \ln_p(N_P(e))\, (P - \bar{P}) \in \bigoplus Q_p (P - \bar{P}),$$

where the sum, as usual, is over $P|p$, one taken from each complex conjugate pair. Here $N_P(e)$ is the local norm $N_{K_P/Q_p}(\phi(e))$, where ϕ is any imbedding for which $P = P_\phi$. Since LOG kills roots of 1 and is linear on E (i.e., $\mathrm{LOG}(e_1e_2) = \mathrm{LOG}(e_1) + \mathrm{LOG}(e_2)$), it extends uniquely to $E \otimes Q$, and so, via Φ, to $\bigoplus Q(P - \bar{P})$:

$$\mathrm{LOG}: \bigoplus Q(P - \bar{P}) \longrightarrow \bigoplus Q_p (P - \bar{P}). \tag{3.5}$$

Since LOG kills only roots of 1 in E, it is easy to see that its image in $\bigoplus Q_p (P - \bar{P})$ has Q-rank g, where $2g$ is the number of primes P over p.

But the interesting question is the Q_p-rank of the image. In other words, are the vectors $\mathrm{LOG}(e_j)$ even Q_p-independent as $\{e_j\}$ runs through a "fundamental set of p-units" (i.e., a maximal set of p-units which are multiplicatively independent)? Gross conjectures that they are.

Gross's first conjecture. _Let_ LOG_{Q_p} _be the endomorphism of_ $\bigoplus Q_p(P-\bar{P})$ _obtained by extending_ LOG^p _linearly from_ $\bigoplus Q(P-\bar{P})$ _to_ $\bigoplus Q_p(P-\bar{P})$. _Define_

$$R_{p,Gross}(K) = \mathrm{Det}\ LOG_{Q_p}.$$

Then

$$R_{p,Gross}(K) \neq 0.$$

We continue to let K denote a CM field, a purely imaginary quadratic extension of the totally real field K^+, and let k denote the totally real ground field. Let $G = \mathrm{Gal}(K/k)$, and let $\rho: G \longrightarrow \mathrm{Aut}(V)$ be a representation in a finite dimensional Ω_p-vector space V. (Note: Any continuous representation $\rho: \mathrm{Gal}(\bar{k}/k) \longrightarrow \mathrm{Aut}(V)$ which is even (resp. odd) factors through $\mathrm{Gal}(K/k)$ for some CM field K, where ρ is even (resp. odd) if its value on complex conjugation is 1 (resp. -1); since we shall only consider such ρ, there is no loss of generality in taking K to be a CM field.)

G also acts on the Ω_p-vector space

$$X \underset{def}{=} \bigoplus \Omega_p(P-\bar{P})$$

by permuting the divisors $P-\bar{P}$. We write P^σ for σP; note that $P^{\sigma\tau} = \sigma\tau P = (P^\tau)^\sigma$. Note that complex conjugation acts by -1. We can combine the action of G on V with its action on X by looking at the subspace

$$(V \underset{\Omega_p}{\otimes} X)^G$$

of G-invariant elements in the tensor product. (In our discussion of $L(1,\chi)$ and the Stark-Tate conjecture in §1, we dealt with $V_\chi \underset{C[G]}{\otimes} X$. For one-dimensional χ, this is isomorphic to $(V_{\bar{\chi}} \otimes X)^G$, reflecting the fact that the behavior of $L(s,\chi)$ near $s=1$ is related by the functional equation to the behavior of $L(s,\bar{\chi})$ near $s=0$.)

To see what $(V\otimes X)^G$ looks like, suppose that G acts transitively on the primes P of K over p (i.e., there is only one prime of k over p), and let P_0 be a fixed prime ideal of K over p. Let $D_0 \subset G$ be the decomposition group of P_0. We shall suppose that the representation ρ is odd. Then as a vector space $(V\otimes X)^G$ is isomorphic to the subspace V^{D_0} of V left fixed by $\rho(D_0)$: namely, let $v \in V^{D_0}$ correspond to

$$\sum_{\sigma \bmod D_0} \rho(\sigma)v \otimes P_0^{\sigma} \in (V\otimes X)^G.$$

To give a simple example, suppose that $k = Q$, $K = Q(\zeta)$, ζ a primitive d-th root of 1, $G \approx (Z/dZ)^*$, and ρ is one-dimensional. Thus, ρ is given by a Dirichlet character $\chi\colon G \longrightarrow \Omega_p^*$. Further suppose that $p = \prod_{\sigma\in G} P_0^{\sigma}$ splits completely in K, i.e., $d|p-1$. In that case $(V\otimes X)^G$ is spanned by the vector $\sum_{\sigma\in G} \chi(\sigma)\, P_0^{\sigma}$. (Compare with the definition (2.4) of $\sigma_\chi \in \Omega_p[G]$ in the proof of Leopoldt's conjecture for K/Q abelian.) More generally, if $p \not\equiv 1 \pmod d$, then the same element spans $(V\otimes X)^G$ if $\chi(p) = 1$, but $(V\otimes X)^G = 0$ if $\chi(p) \neq 1$.

Returning to the general case of an odd representation ρ of $G = \mathrm{Gal}(K/k)$, K a CM field and k totally real, we see that the endomorphism LOG_{Ω_p} of $X = \bigoplus \Omega_p (P - \bar{P})$ is G-equivariant. This is because, for $e \in E = E^{(p)}(K)$ and $\sigma \in G$,

$$\mathrm{LOG}(\sigma e) = \sum_{\phi} \ln_p \phi(\sigma e)\, P_\phi = \sum_{\phi} \ln_p \phi(e)\, P_{\phi\sigma^{-1}} = \sum_{\phi} \ln_p \phi(e)\, P_\phi^{\sigma},$$

since we have $P_{\phi\sigma^{-1}} = P_\phi^{\sigma} = \sigma P_\phi$ directly from the definition (3.4).

Thus, LOG_{Ω_p} induces an endomorphism of $(V\otimes X)^G$, which we denote LOG_V. Gross's regulator for ρ is defined as

$$R_{p,\mathrm{Gross}}(\rho) = \mathrm{Det}\ \mathrm{LOG}_V .$$

If the first Gross conjecture is true, then LOG_V is also an isomorphism, and $R_{p,Gross}(\rho) \neq 0$.

The earlier regulator $R_{p,Gross}(K)$ is a special case of $R_{p,Gross}(\rho)$. Namely, let $k = K^+$, and let $\rho\colon Gal(K/K^+) \longrightarrow \{\pm 1\}$ be the unique nontrivial character. Then $(V \otimes X)^G \approx X$, and $R_{p,Gross}(\rho) = R_{p,Gross}(K)$.

Recall that if $G = Gal(K/k)$ permutes the primes P of K lying over p --i.e., if there is only one prime of k lying over p --then $\dim (V \otimes X)^G = \dim V^{D_0}$, where D_0 is one of the decomposition groups. More generally, if there is more than one prime of the ground field k lying over p, we have the picture

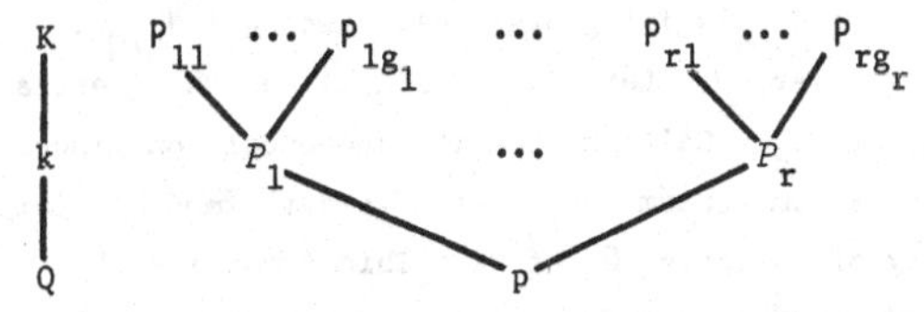

Then for each i, G permutes the P_{ij}, $j = 1,\ldots,g_i$, lying over $\mathcal{P}_i$, and the same argument shows that $(V \otimes X)^G$ is isomorphic as a vector space to $\bigoplus V^{D_i}$, where D_i is the decomposition group of P_{i1}. We let

$$m_\rho = \dim (V \otimes X)^G = \sum_i \dim V^{D_i}.$$

Then $R_{p,Gross}(\rho)$ is the determinant of an $m_\rho \times m_\rho$ matrix.

For example, if V is one-dimensional, i.e., if $\rho\colon Gal(K/k) \longrightarrow \Omega_p^*$ is a one-dimensional character, then m_ρ is equal to the number of primes $\mathcal{P}_i$ of k lying over p such that ρ is trivial on one (and hence on all) of the decomposition groups D_i of primes of K lying over $\mathcal{P}_i$. In the case of one-dimensional ρ, conjecturally the m_ρ vanishing Euler factors $(1 - \rho(\mathcal{P}_i))$ in the Deligne-Ribet [21] function $L_p(s,\rho\omega)$ at $s = 0$ should lead to an m_ρ-fold zero at $s = 0$; and it is further conjectured [29] that the zero is of order exactly m_ρ. But it has not even been proved that $L_p(s,\rho\omega)$ has a multiple zero at $s = 0$ when $m_\rho > 1$.

Gross's conjecture, which presumes that the order of zero is at least m_ρ, concerns the coefficient of s^{m_ρ} in the Taylor expansion at $s = 0$ of $L_p(s,\rho\omega)$. In terms of his conjecture, we shall see that the assertion that $L_p(s,\rho\omega)$ has exactly an m_ρ-fold zero at $s = 0$ is equivalent to non-vanishing of $R_{p,\text{Gross}}(\rho)$.

Gross's conjectured leading coefficient of $L_p(s,\rho\omega)$ at $s = 0$ includes a certain algebraic number coming from the complex-analytic Artin L-series whose special values are p-adically interpolated by $L_p(s,\rho)$. We first recall the definition of the Artin L-series associated to a representation ρ: $\text{Gal}(K/k) \longrightarrow \text{Aut}(V)$, where now V is a finite dimensional complex vector space. Let P be any prime of k (not necessarily lying over p). Let $q = N_{k/Q}P$. Let P be a prime of K over P. Let $I_{\mathrm{P}} \subset \text{Gal}(K/k)$ be the inertia group of P, and let $D_{\mathrm{P}} \subset \text{Gal}(K/k)$ be its decomposition group. Let $F_{\mathrm{P}} \in D_{\mathrm{P}}$ be any automorphism such that $F_{\mathrm{P}}x \equiv x^q \pmod{\mathrm{P}}$ for any x in the ring of integers $\mathcal{O}$ of K. This Frobenius F_{P} is uniquely determined up to an element of I_{P}. Hence the "local factor at P"

$$\text{Det}\left(1 - q^{-s}\rho(F_{\mathrm{P}})|V^{I_{\mathrm{P}}}\right),$$

where $V^{I_{\mathrm{P}}}$ is the subspace of vectors fixed by $\rho(I_{\mathrm{P}})$, does not depend on the choice of Frobenius for P. If we change the choice of prime P of K lying over P, the effect is to conjugate F_{P} and I_{P} by some element of $\text{Gal}(K/k)$. Hence the determinant is unaffected. Thus, the above local factor depends only on ρ, P, and the complex variable s. The Artin L-series $L(s,\rho)$ is defined as the product of these local factors over all primes P of k. This Euler product converges for $\text{Re}(s) > 1$ and has a meromorphic continuation onto the entire complex plane. The Artin conjecture asserts that it is holomorphic, except for a pole at $s = 1$ if ρ contains the trivial representation.

Example. Take the simple case when $k = Q$, $K = Q(\zeta)$, ζ a primitive d-th root of 1, $G = \text{Gal}(K/k) \approx (Z/dZ)^*$, and ρ is a primitive character χ: $G \longrightarrow C^*$. If the ideal (p) of Q divides d, then it ramifies in K, and I_P (for any $P|p$) is the kernel of the map $(Z/dZ)^* \longrightarrow (Z/d'Z)^*$, where $d' = d/p^{\mathrm{ord}_p d}$. Since χ is primitive, it is nontrivial on I_P, and so $V^{I_P} = 0$. Thus, the local factor at p is 1.

If, on the other hand, $p \nmid d$, then $F_P = p \in (Z/dZ)^*$, and the local factor is $(1 - p^{-s}\chi(p))^{-1}$. Hence,

$$L(s,\rho) = \prod_{p \nmid d} (1 - p^{-s}\chi(p))^{-1} = \sum \frac{\chi(n)}{n^s} = L(s,\chi),$$

which is the usual Dirichlet L-series.

Recall that in the case of Dirichlet L-series, before p-adically interpolating its values at negative integers we had to modify it by removing the Euler factor at p:

$$L^*(s,\chi) = \sum_{p \nmid n} \frac{\chi(n)}{n^s} = \prod_{\ell \neq p} (1 - \ell^{-s}\chi(\ell))^{-1} = (1 - p^{-s}\chi(p))L(s,\chi).$$

A similar modification is required in the general case of Artin L-series before we can make the transition to p-adic Artin L-series. Namely, given our fixed prime p, we define the modified Artin L-series $L^*(s,\rho)$ to be the product of the local factors at all primes of k <u>not dividing</u> p.

In the case of Dirichlet L-series, the values at negative integers $1-n$ are $-B_{n,\chi}/n \in Q(\chi)$, the field generated by the values of χ. A similar fact was proved for Artin L-series $L(s,\rho)$ by Siegel [89]. Namely, first note that the representation ρ: Gal(K/k) $\longrightarrow$ Aut(V) can be obtained by extension of scalars from a representation in a $\mathcal{K}$-vector space $V_{\mathcal{K}}$, where $\mathcal{K}$ is a finite Galois extension of Q. (In other words, for a suitable basis of V, the matrix extries in $\rho(\sigma)$, $\sigma \in \mathrm{Gal}(K/k)$, are all in $\mathcal{K}$.) Then Siegel showed that $L(1-n,\rho) \in \mathcal{K}$, and, if $\sigma\rho$, $\sigma \in \mathrm{Gal}(\mathcal{K}/Q)$, denotes the representation obtained by composing ρ with the action of σ on $\mathrm{Aut}(V_{\mathcal{K}})$, then $L(1-n,\sigma\rho) = \sigma L(1-n,\rho)$. The same is then clearly true of the modified L-function $L^*(1-n,\rho)$.

By fixing once and for all an imbedding $\overline{Q} \hookrightarrow \Omega_p$, we can consider ρ as giving a p-adic representation, which we also denote ρ, in $V_K \otimes_K \Omega_p$, which we also denote V. Then there exists a meromorphic function $L_p(s,\rho)$ on Z_p with values in Ω_p which satisfies

$$L_p(1-n,\rho) = L^*(1-n,\rho \otimes \omega^{-n}), \quad n \geq 2$$

(where we use the fixed imbedding $\overline{Q} \hookrightarrow \Omega_p$ to identify complex and p-adic representations and L-function values). When $n = 1$, this relation

$$L_p(0,\rho) = L^*(0,\rho \otimes \omega^{-1})$$

is also known to hold if ρ is one-dimensional (or a direct sum of representations induced from one-dimensional representations); it is conjectured to hold for general ρ. $L_p(s,\rho)$ is identically zero unless ρ is an even representation.

Gross now defines a second modification $L^{**}(s,\rho)$ of the Artin L-series. Recall that to get the first modification $L^*(s,\rho)$, we threw out the local factors at primes P of k over p, by multiplying by the determinants (here $\mathfrak{P}$ is any prime of K over P, $q = NP$)

$$\mathrm{Det}\left(1 - q^{-s}\rho(F_{\mathfrak{P}})|V^{I_{\mathfrak{P}}}\right).$$

To get $L^{**}(s,\rho)$, we put back in part of that local factor, by dividing by the subdeterminants

$$\mathrm{Det}\left(1 - q^{-s}\rho(F_{\mathfrak{P}})|V^{D_{\mathfrak{P}}}\right)$$

for $P|p$, where we restrict $F_{\mathfrak{P}}$ to the part of V invariant under the whole decomposition group $D_{\mathfrak{P}}$. Since $F_{\mathfrak{P}} \in D_{\mathfrak{P}}$, we have $\rho(F_{\mathfrak{P}})|V^{D_{\mathfrak{P}}} = \text{identity}$, and so we define

$$L^{**}(s,\rho) \underset{\mathrm{def}}{=} L^*(s,\rho) \prod_{P|p} (1 - q^{-s})^{-\dim V^{D_{\mathfrak{P}}}}.$$

For example, if ρ is one-dimensional, this means that we put back in the Euler factors $(1 - q^{-s}\rho(F_{\mathfrak{P}}))^{-1}$ when $\rho(F_{\mathfrak{P}}) = 1$. For instance, if $k = Q$, $K = Q(\zeta)$, ζ a primitive d-th root of 1, $G = \mathrm{Gal}(K/k) \approx (Z/dZ)^*$, and ρ corresponds to the Dirichlet character χ, then $L^*(s,\chi) = (1 - p^{-s}\chi(p))\,L(s,\chi)$, and

$$L^{**}(s,\chi) = \begin{cases} L^*(s,\chi) & \text{if } \chi(p) \neq 1, \\ L(s,\chi) & \text{if } \chi(p) = 1. \end{cases}$$

The reason for this second modification is as follows. The subdeterminants $\mathrm{Det}\left(1 - q^{-s}\,\rho(F_P)\,|\,V^{D_P}\right)$ of the factors $\mathrm{Det}\left(1 - q^{-s}\,\rho(F_P)\,|\,V^{I_P}\right)$ that are thrown in to get $L^*(s,\rho)$ bring in zeros at $s = 0$ of order $\dim V^{D_P}$. Hence, $L^*(s,\rho)$ has an m_ρ-fold zero at $s = 0$, where $m_\rho = \sum_P \dim V^{D_P}$. Since $L_p(1-n,\rho\otimes\omega^n)$ interpolates the values $L^*(1-n,\rho)$, it is conjectured that $L_p(s,\rho\omega)$ also has an m_ρ-fold zero at $s = 0$, but this by no means follows from the mere fact that $L_p(s,\rho)$ interpolates $L^*(s,\rho)$.

To obtain the coefficient of the leading term of $L_p(s,\rho)$ at $s = 0$, Gross therefore divides by the factors that conjecturally give the zeros at $s = 0$.

Thus, the function whose value at $s = 0$ is conjecturally related to this leading term is

$$L^{**}(s,\rho) = L(s,\rho) \prod \mathrm{Det}\left(1 - q^{-s}\,\rho(F_P)\,|\,V^{I_P}/V^{D_P}\right),$$

where, as usual, the product is over all primes P of k over p, $q = NP$, and P is some fixed choice of prime of K over P for each P.

Since $L_p(s,\rho)$ is only a nonzero function when ρ is even, and since its value at $s = 0$ is related to $L(s,\rho\otimes\omega^{-1})$, if we replace ρ by $\rho\otimes\omega$ we see that $L_p(s,\rho\otimes\omega)$ at $s = 0$ should be related to $L(s,\rho)$, or rather $L^{**}(s,\rho)$, for ρ an <u>odd</u> representation.

For an odd representation ρ, Gross defines (A stands for "algebraic part"):

$$A(\rho) = L^{**}(0,\rho) \prod_P f_P^{\dim V^{D_P}},$$

where the product is over all primes P of k over p, D_P is the decomposition group of a prime P of K over $\mathcal{P}$, and $f_{\mathcal{P}}$ is the residue degree $[o/\mathcal{P}: F_p]$, where o is the ring of integers of k ($f_{\mathcal{P}}$ should not be confused with the residue degree $[O/P: F_p]$ of P, or with the relative residue degree $[O/P: o/\mathcal{P}] = \# D_P$; in any case, this product term is just 1 if $k = Q$.)

Without further ado, we can finally state Gross's main conjecture.

Gross's second conjecture. _If_ ρ: $\mathrm{Gal}(K/k) \longrightarrow \mathrm{Aut}(V)$ _is an odd representation in a finite dimensional_ Ω_p_-vector space_ V, _then_ $L_p(s, \rho \otimes \omega)$ _has a zero of order exactly_ $m_\rho = \Sigma \dim V^{D_P}$ _at_ $s = 0$, _and_

$$\lim_{s \to 0} s^{-m_\rho} L_p(s, \rho \otimes \omega) = R_{p,\mathrm{Gross}}(\rho) A(\rho).$$

4. Gross's conjecture in the abelian over Q case

We now prove this conjecture when $k = Q$ and $G = \mathrm{Gal}(K/Q)$ is abelian. The conjecture is unproved in essentially any other case, even, for example, when $k = Q(\sqrt{D})$ and K/k is abelian. Without loss of generality we may suppose that $K = Q(\zeta)$, ζ a primitive d-th root of 1, since any abelian extension of Q is contained in such a K, and all of the expressions in the conjecture remain the same if K is replaced by a larger field. We first prove:

Proposition. _Gross's first conjecture holds in the abelian over_ Q _case, i.e.,_ LOG_{Q_p} _is an automorphism of_ $\bigoplus Q_p (P - \bar{P})$.

Proof. It suffices to prove that LOG_{Q_p} is an automorphism when we extend scalars from Q_p to Ω_p. (We want to go to an algebraically closed field, so that we can decompose by the action of characters of $G = \mathrm{Gal}(K/Q) \approx (Z/dZ)^*$.) Thus, we shall show that

$$\begin{aligned} \mathrm{LOG}: \quad E \otimes \Omega_p &\longrightarrow \bigoplus \Omega_p (P - \bar{P}) \\ e &\longmapsto \sum_P \ln_p(N_P(e)) \, (P - \bar{P}) \end{aligned} \tag{4.1}$$

is an isomorphism, where the sum is over primes P of K over p, one from each complex conjugate pair. (Recall that $N_P(e)$ denotes $N_{K_P/Q_p}(\phi(e))$, where ϕ is any imbedding $K \hookrightarrow \Omega_p$ for which $P = \{x \in O \mid |\phi(x)|_p < 1\}$.)

G acts on both $E \otimes \Omega_p$ and $\oplus \Omega_p (P - \bar{P})$, and we have seen that LOG is G-equivariant. Let us decompose both sides by characters χ of G. It is easy to see that the χ-component of each side is at most one-dimensional; it is nonzero if and only if χ is odd and $\chi(p) = 1$. In that case the χ-component $(E \otimes \Omega_p)^\chi$ is spanned by $e_\chi \overset{=}{\text{def}} \Sigma \bar{\chi}(n)\sigma_n(e)$, where $e = \alpha/\bar{\alpha}$ is a p-unit with $(\alpha) \doteq P^h$; and $(\oplus \Omega_p (P - \bar{P}))^\chi$ is spanned by $\Sigma \bar{\chi}(n) P^{\sigma_n}$, where P is any fixed prime ideal of K over p. It therefore suffices to show that $\mathrm{LOG}(e_\chi) \neq 0$.

If we denote $G = (Z/dZ)^*$, $D = D_P = \{p^j\} \subset G$, and $f = \#D$, then by (4.1) we have

$$\begin{aligned}\mathrm{LOG}(e_\chi) &= 2f \sum_{m \in G/\pm D} \sum_{n \in G/\pm D} \bar{\chi}(m) \ln_p N_{P^{\sigma_n}}(\sigma_m(e)) (P^{\sigma_n} - \bar{P}^{\sigma_n}) \\ &= 2f \sum_{n,j \in G/\pm D} \bar{\chi}(n) \chi(j) \ln_p N_{P^{\sigma_j}}(e) (P^{\sigma_n} - \bar{P}^{\sigma_n}) \qquad (j = \tfrac{n}{m}) \\ &= \Big(\sum_{n \in G/\pm D} \chi(n) \ln_p N_{P^{\sigma_n}}(e)\Big) \Big(\sum_{n \in G} \bar{\chi}(n) (P^{\sigma_n} - \bar{P}^{\sigma_n})\Big).\end{aligned}$$

If this is zero, then from the p-adic Baker theorem (see §2) it follows that $\ln_p$ of the algebraic numbers $N_{P^{\sigma_n}}(e)$ must be dependent over Q. For brevity let e_n denote the product $\Pi\, \sigma_n^{-1}(\tau e)$ taken over all $\tau \in D$, so that $N_{P^{\sigma_n}}(e) = N_P(\sigma_n^{-1}(e)) = \phi(e_n)$ for ϕ any fixed imbedding $K \hookrightarrow \Omega_p$ for which $P = \{x \in O \mid |\phi(x)|_p < 1\}$. Thus, for some $m_n \in Z$ not all zero we have

$$\ln_p \Big(\phi\Big(\prod_{n \in G/\pm D} e_n^{m_n}\Big)\Big) = 0.$$

This means that $\Pi\, e_n^{m_n}$ must be a power of p times a root of unity; replacing m_n by a suitable multiple, we obtain $\Pi\, e_n^{m_n} = p^r$ for some $r \in Z$. But the ideal decomposition of $\Pi\, e_n^{m_n}$ is

$$\left(\prod_{n \in G/\pm D} \left(p^{\sigma_n^{-1}}/\bar{p}^{\sigma_n^{-1}}\right)^{m_n}\right)^{hf},$$

while $(p)^r = \left(\prod_{n \in G/D} P^{\sigma_n}\right)^r$. This contradiction proves the proposition.

It is curious to note the resemblance between this proof and the proof of Leopoldt's conjecture for K/Q abelian: in both cases the key step was to use the p-adic Baker theorem to conclude from the vanishing of a character sum that certain units are multiplicatively dependent.

Theorem. *Gross's second conjecture holds in the abelian over Q case.*

Proof. Again $k = Q$, $K = Q(\zeta)$, ζ a primitive d-th root of 1, $G = \mathrm{Gal}(K/Q) \approx (Z/dZ)^*$. Since any representation of the abelian group G decomposes into a direct sum of one-dimensional characters, and since both sides of Gross's conjecture can readily be verified to be multiplicative with respect to direct sums of representations, we can reduce to the case when ρ is one-dimensional, i.e., is a Dirichlet character $\rho: G = (Z/dZ)^* \longrightarrow \bar{Q}^*$. (We shall continue to use the letter ρ for this Dirichlet character, rather than χ, since the letter χ will soon be needed to denote a completely different kind of character, namely, a character of the multiplicative group of a finite field.)

Now there are two cases, depending on whether or not ρ is trivial on the decomposition group $D = D_P = \{p^j\} \subset G$ (P a prime of K over p), i.e., we must consider: case (i) $\rho(p) \neq 1$, $m_\rho = \dim V^D = 0$; and case (ii) $\rho(p) = 1$, $m_\rho = 1$. Without loss of generality we may suppose that d is the conductor of ρ; otherwise, ρ factors through $\mathrm{Gal}(Q(\zeta')/Q)$, where ζ' is a primitive $(\mathrm{cond}\ \rho)$-th root of 1.

Case (i) $m_\rho = 0$

In this case $R_{p,\mathrm{Gross}}(\rho) = 1$, $A(\rho) = L^{**}(0,\rho) = L^*(0,\rho) =$

$= (1-\rho(p))\, L(0,\rho)$, and Gross's conjecture says that

$$L_p(0,\rho\omega) = L(0,\rho)\ (1-\rho(p)),$$

which is true (see §II.4); in fact, both sides equal $-(1-\rho(p))B_{1,\rho}$. We also know that this expression is nonzero (since ρ is odd and primitive, and $\rho(p) \neq 1$), in other words, the order of vanishing of $L_p(s,\rho\omega)$ at $s=0$ is $m_\rho = 0$.

Case (ii) $m_\rho = 1$, i.e., $\rho(p) = 1$

This is the more interesting case.

We first compute $R_{p,\mathrm{Gross}}(\rho)$. Let $\phi\colon K \hookrightarrow \Omega_p$ be a fixed imbedding, so that $P = \{x \in O \mid |\phi(x)|_p < 1\}$ denotes a fixed prime of K over p. As before, let $D = D_p = \{p^j\} \subset G = (Z/dZ)^*$, and let $f = \#D$. The one-dimensional vector space $(V \otimes X)^G$ is spanned by

$$\sum_{\sigma \in G/D} \rho(\sigma)\, P^\sigma \in X = \bigoplus_{\sigma \in G/\pm D} \Omega_p\ (P^\sigma - \bar{P}^\sigma).$$

Recall how LOG was defined on such an element. Let $P^h = (\alpha)$. Since $((\alpha/\bar{\alpha})^\sigma) = (P/\bar{P})^h$ for $\sigma \in D$, it follows that $\alpha/\bar{\alpha}$ and $(\alpha/\bar{\alpha})^\sigma$ for $\sigma \in D$ differ by a root of unity (since their quotient is a unit of K with complex absolute value 1). The image of the divisor $\sum_{\sigma \in G/D} \rho(\sigma)\, P^\sigma$ under the isomorphism Φ is (see (3.3))

$$-\frac{1}{hf} \sum_{\sigma \in G/\pm D} \rho(\sigma)\ (\alpha/\bar{\alpha})^\sigma \in E \otimes \Omega_p,$$

and, since

$$N_{p^\tau}((\alpha/\bar{\alpha})^\sigma) = N_{K_{p^\tau}/Q_p}(\phi\circ\tau^{-1}(\alpha/\bar{\alpha})^\sigma) = \phi\Big(((\alpha/\bar{\alpha})^{\sigma\tau^{-1}})^f \cdot \text{root of } 1\Big),$$

it follows that

$$\mathrm{LOG}\Big(\sum_{\sigma \in G/D} \rho(\sigma)\, P^\sigma\Big) = -\frac{1}{hf} \sum_{\sigma \in G/\pm D} \rho(\sigma) \sum_{\tau \in G/D} \ln_p N_{p^\tau}((\alpha/\bar{\alpha})^\sigma)\ P^\tau \quad \text{(see (3.5))}$$

$$= -\frac{1}{h} \sum_{\sigma \in G/\pm D,\ \tau \in G/D} \rho(\sigma)\ \ln_p \phi\left((\alpha/\bar{\alpha})^{\sigma\tau^{-1}}\right) P^\tau$$

$$= -\frac{1}{h}\Big(\sum_{\tau \in G/D} \rho(\tau)\, P^\tau\Big)\Big(\sum_{\sigma \in G/\pm D} \rho(\sigma) \ln_p \phi((\alpha/\bar{\alpha})^\sigma)\Big).$$

Hence

$$R_{p,\text{Gross}}(\rho) = -\frac{1}{h} \sum_{\sigma \in G/D} \rho(\sigma) \ln_p \phi(\alpha^\sigma),$$

where α is any generator of the ideal P^h.

Gross's conjecture asserts that

$$L_p'(0,\rho\omega) = R_{p,\text{Gross}}(\rho)\, A(\rho),$$

where $A(\rho) = L(0,\rho) = -B_{1,\rho}$.

We now use the following results from earlier chapters:

(1) the formula for $L_p'(0,\rho\omega)$ in §II.8

$$L_p'(0,\rho\omega) = \sum_{0<a<d} \rho(a) \ln_p \Gamma_p(a/d);$$

(2) the p-adic formula for Gauss sums in §III.6, which, after we take $\ln_p$ and use the fact that $\ln_p \pi = \frac{1}{p-1} \ln_p(-p) = 0$, gives

$$\ln_p g(\tilde{\chi}_a^{-1}, \psi_\pi \circ \text{Tr}) = \sum_{b \in D} \ln_p \Gamma_p(\langle ab/d \rangle)$$

(see §III.6 for notation; here $\tilde{\chi}_a$ is a multiplicative character of a finite field);

(3) Stickleberger's theorem (§III.7), which tells us that the ideal decomposition of $g(\tilde{\chi}_a^{-1}, \psi_\pi \circ \text{Tr})^d$, written additively, is $\sum_{j \in G} \langle a/j \rangle_d P^{\sigma_j}$, where $\langle\ \rangle_d$ denotes least positive residue mod d.

We conclude that

$$\begin{aligned}
L_p'(0,\rho\omega) &= \sum_{a \in G/D} \rho(a) \sum_{b \in D} \ln_p \Gamma_p(\langle ab/d \rangle) \\
&= \sum_{a \in G/D} \rho(a) \ln_p g(\tilde{\chi}_a^{-1}, \psi_\pi \circ \text{Tr}) \\
&= \frac{1}{dh} \sum_{a \in G/D} \rho(a) \sum_{j \in G} \langle a/j \rangle_d \ln_p \phi(\alpha^{\sigma_j}) \\
&\qquad\qquad \text{(where } P^h = (\alpha) \text{ as before)} \\
&= \frac{1}{dhf} \sum_{a,j \in G} \rho(a/j) \langle a/j \rangle_d\, \rho(j) \ln_p \phi(\alpha^{\sigma_j}) \\
&= \frac{1}{dhf} \sum_{0<a<d} a\, \rho(a) \sum_{\sigma \in G} \rho(\sigma) \ln_p \phi(\alpha^\sigma) \\
&= \left(-\frac{1}{d} \sum_{0<a<d} a\, \rho(a)\right)\left(-\frac{1}{h} \sum_{\sigma \in G/D} \rho(\sigma) \ln_p \phi(\alpha^\sigma)\right),
\end{aligned}$$

which is precisely $(-B_{1,\rho})\ R_{p,\mathrm{Gross}}(\rho)$, as desired. Q.E.D.

The above proof of Gross's first and second conjectures in the abelian over $\mathbb{Q}$ case is Gross's variant of Ferrero-Greenberg's original proof in [29] of the simplicity of the zero of $L_p(s,\rho\omega)$ at $s=0$. The proof relies in an essential way upon several basic and diverse p-adic results: Brumer's p-adic Baker theorem, the formula relating $L_p'(0,\rho\omega)$ to special values of the p-adic gamma function, Stickleberger's theorem, and the p-adic formula for Gauss sums. With this proof we conclude the main part of the book.

APPENDIX

1. A theorem of Amice-Frèsnel

The measure μ_z on Z_p defined by $\mu_z(a + p^N Z_p) = z^a/(1-z^{p^N})$, which was used to study p-adic Dirichlet L-functions in Chapter II, can also be used to give a simple proof of the following general fact.

Theorem (Amice and Frèsnel [4]). *Let $f(z) = \Sigma a_n z^n \in \Omega_p[[z]]$ have the property that the coefficients a_n can be p-adically interpolated, i.e., there exists a continuous function $\phi: Z_p \longrightarrow \Omega_p$ such that $\phi(n) = a_n$. Then f (whose disc of convergence must be the open unit disc $D_0(1^-)$) is the restriction to $D_0(1^-)$ of a Krasner analytic function $\tilde{f}$ on the complement of $D_1(1^-)$. In addition, $\tilde{f}$ has the Taylor series at infinity*

$$\tilde{f}(z) = - \sum_{n=1}^{\infty} \phi(-n)\, z^{-n}, \qquad |z|_p > 1.$$

Proof. Define

$$\tilde{f}(z) = \int \phi \, d\mu_z .$$

Then on $\Omega_p - D_1(1^-)$, the function $\tilde{f}$ is the uniform limit as $N \longrightarrow \infty$ of the following rational functions with poles in $D_1(1^-)$:

$$\sum_{0 \le n < p^N} \phi(n) \frac{z^n}{1 - z^{p^N}}$$

(see (2.3) of §II.2). Hence $\tilde{f}$ is Krasner analytic on $\Omega_p - D_1(1^-)$.
Next, for $|z|_p < 1$ we have $\lim_{N \to \infty} 1/(1-z^{p^N}) \doteq 1$, and so

$$\tilde{f}(z) = \lim_{N \to \infty} \sum_{0 \le n < p^N} \phi(n) z^n = f(z).$$

Finally, if $|z|_p > 1$, then we have

$$\sum_{0 \le n < p^N} \phi(n) \frac{z^n}{1 - z^{p^N}} = \sum_{0 < n \le p^N} \phi(p^N - n) \frac{z^{-n}}{z^{-p^N} - 1}$$

$$\longrightarrow - \sum_{n=1}^{\infty} \phi(-n) z^{-n}$$

as $N \longrightarrow \infty$, and the theorem is proved.

2. The classical Stieltjes transform

The Stieltjes transform of a function $f: [0,\infty) \longrightarrow C$ is

$$G(z) = \int_0^\infty \frac{f(x)}{x+z} dx \tag{2.1}$$

for all z such that the integral converges. (More generally, one can replace $f(x)dx$ by $dF(x)$ and define the transform of that measure to be the corresponding Stieltjes integral.) Usually $f(x)$ is a rapidly decreasing function, and the integral converges for all $z \in C - (-\infty,0]$. The Stieltjes transform is the square of the Laplace transform L:

$$L(L(f))(z) = \int_0^\infty e^{-zt}\left(\int_0^\infty e^{-tx} f(x)dx\right)dt$$

$$= \int_0^\infty f(x) \int_0^\infty e^{-(z+x)t} dt\, dx = \int_0^\infty \frac{f(x)}{x+z} dx.$$

It has been used extensively in the study of continued fraction expansions of analytic functions (this was Stieltjes' original purpose), numerical analysis, and quantum mechanics.

The function f need not be rapidly decreasing in order for the Stieltjes transform (2.1) to exist. The Stieltjes transform also converges for $z \notin (-\infty,0]$ if $f: [0,\infty) \longrightarrow C$ is a periodic function satisfying the conditions

$$(1) \quad f(x+d) = f(x); \tag{2.2}$$

$$(2) \quad f \in L^1([0,d]); \tag{2.3}$$

$$(3) \quad \int_0^d f(x)\,dx = 0. \tag{2.4}$$

This is because

$$|G(z)| = \left| \sum_{n=1}^{\infty} \int_{(n-1)d}^{nd} \left(\frac{f(x)}{x+z} - \frac{f(x)}{nd} \right) dx \right| \qquad \text{by (2.4)}$$

$$\leq \sum_{n=1}^{\infty} \frac{\text{const}}{(nd)^2} \int_0^d |f(x)|\,dx < \infty,$$

where "const" depends on z but not on n. It is periodic f that we shall be particularly interested in from a number theoretic point of view.

We may suppose that in addition to (2.2)-(2.4) the function $f: [0,\infty) \longrightarrow C$ satisfies

$$(4) \quad f(x) = 0 \text{ for } x < \delta \text{ for some positive } \delta. \tag{2.5}$$

Example. Let χ be a nontrivial <u>even</u> (i.e., $\chi(-1) = 1$) Dirichlet character of conductor d. Define

$$f_\chi(x) = \sum_{a=1}^{[x]} \chi(a). \tag{2.6}$$

Then f_χ obviously satisfies (2.2), (2.3) and (2.5) (with $\delta = 1$). To verify (2.4), we compute

$$\int_0^d f_\chi(x)\,dx = \sum_{a=1}^{d} (d-a)\chi(a) = -\sum_{a=1}^{d} a\chi(a) = -dB_{1,\chi} = 0$$

for χ even. (If χ were odd, we would have to add the constant $B_{1,\chi}$ to f_χ, and (2.5) would no longer hold; for simplicity in the discussion below, we want to assume (2.5).)

Suppose that $f: [0,\infty) \longrightarrow C$ satisfies (2.2)-(2.5). Then

(2.1) converges for $z \notin (-\infty, -\delta]$. For $|z| < \delta$ we expand

$$G(z) = \int_0^\infty \frac{f(x)}{x+z}\,dx$$

$$= \sum_{n=0}^\infty (-z)^n \int_0^\infty f(x)x^{-n-1}dx. \qquad (2.7)$$

Note that the "negative moments" $\int_0^\infty f(x)x^{-n-1}dx$ are convergent integrals, and are easily seen to be $O(\delta^{-n})$ as $n \longrightarrow \infty$. So (2.7) is the Taylor series of $G(z)$ in $|z| < \delta$.

In our example f_χ, we compute for $s > 0$

$$\int_1^\infty f_\chi(x)x^{-s-1}dx = \sum_{k=1}^\infty \frac{1}{s}\left(\frac{1}{k^s} - \frac{1}{(k+1)^s}\right)\sum_{a=1}^k \chi(a)$$

$$= \frac{1}{s}\sum_{k=1}^\infty \frac{\chi(k)}{k^s} = \frac{L(s,\chi)}{s}.$$

Thus for $f = f_\chi$ we have

$$G(z) = \lim_{s \to 0} \frac{L(s,\chi)}{s} + \sum_{n=1}^\infty \frac{L(n,\chi)}{n}(-z)^n, \quad |z| < 1. \qquad (2.8)$$

Returning to the general situation, suppose that $f: [0,\infty) \longrightarrow \mathbb{C}$ satisfies (2.2)-(2.4) (not necessarily (2.5)). Define $f^{(-1)}$ to be the integral of f with constant of integration chosen so that $\int_0^d f^{(-1)}(x)dx = 0$; thus, if f is represented by the Fourier series $\sum a_n e^{2\pi inx/d}$, with $a_0 = 0$ because of (2.4), then $f^{(-1)}$ has Fourier series $\frac{d}{2\pi i}\sum_{n\neq 0} \frac{a_n}{n} e^{2\pi inx/d}$. Then define $f^{(-j-1)}$ inductively as $\left(f^{(-j)}\right)^{(-1)}$, $j = 1, 2, \ldots$.

We obtain an asymptotic series for $G(z)$ as $z \longrightarrow \infty$ (along any ray other than the negative real axis) by integrating by parts:

$$G(z) = \int_0^\infty \frac{f(x)}{x+z}\,dx = -\frac{f^{(-1)}(0)}{z} + \int_0^\infty \frac{f^{(-1)}(x)}{(x+z)^2}\,dx = \cdots$$

$$= -\frac{f^{(-1)}(0)}{z} - \frac{f^{(-2)}(0)}{z^2} - \cdots - \frac{(j-1)!\,f^{(-j)}(0)}{z^j} - \cdots$$

$$- \frac{(n-1)!\,f^{(-n)}(0)}{z^n} + n!\int_0^\infty \frac{f^{(-n)}(x)}{(x+z)^{n+1}}\,dx$$

for any n. For fixed n, if z approaches infinity away from the negative real axis, then it is easy to see that the integral term is $O(z^{-n-1})$. Thus, $-\sum (n-1)!\,f^{(-n)}(0)\,z^{-n}$ is an asymptotic series for $G(z)$.

For example, if $f = f_\chi$ (χ a nontrivial even character), we have the easily computed Fourier series expansion

$$f_\chi(x) = \sum_{a=1}^{[x]} \chi(a)$$

$$= \sum_{n\neq 0} a_n\, e^{2\pi i n z/d}, \quad \text{where} \quad a_n = \frac{g_\chi}{2\pi i}\,\frac{\overline{\chi}(n)}{n}$$

(here $g_\chi = \sum_{0<a<d} \chi(a)\, e^{2\pi i a/d}$ is the Gauss sum for χ). Then

$$f^{(-j)}(0) = \left(\frac{d}{2\pi i}\right)^j \sum_{n\neq 0} \frac{a_n}{n^j} = \frac{d^j g_\chi}{(2\pi i)^{j+1}} \sum_{n\neq 0} \frac{\overline{\chi}(n)}{n^{j+1}}$$

$$= \begin{cases} \dfrac{d^j g_\chi}{(2\pi i)^{j+1}}\, 2L(j+1,\overline{\chi}), & j \text{ odd;} \\ 0, & j \text{ even} \end{cases}$$

$$= \frac{1}{j!}\, L(-j,\chi)$$

by the functional equation for $L(s,\chi)$ (see, e.g., [41], p. 104).

Thus, for $f = f_\chi$ we have

$$G(z) \sim -\sum_{j=1}^{n} \frac{L(-j,\chi)}{j} z^{-j} = -\sum_{j=1}^{n} \frac{L(-j,\chi)}{-j} (-z)^{-j} \qquad (2.9)$$

(since $L(-j,\chi) = 0$ for j even).

Comparing (2.8) and (2.9), we see that we have established a special case of the following theorem of Mellin-LeRoy (see [65], p. 109, 113).

Theorem. *Suppose* $\phi(s)$ *is a function which is holomorphic and bounded on* $\mathrm{Re}\ s \geq 0$. *Then the Taylor series* $G(z) = \sum \phi(n) z^n$ *extends holomorphically onto* $\mathbb{C} - [0,\infty)$ *and has asymptotic series* $G(z) \sim -\Sigma\,\phi(-n) z^{-n}$ *as* $z \longrightarrow \infty$ *(along any ray other than the positive real axis).*

In our example $\phi(s) = \frac{1}{s} L(s,\chi)$ (which is bounded even as $s \longrightarrow 0$, since $L(0,\chi) = 0$ for χ even and nontrivial), and we have replaced z by $-z$.

Remarks. 1. This theorem is the classical analog of the p-adic theorem of Amice-Frèsnel in §1.

2. Under a weaker assumption on $\phi(s)$, namely, bounded exponential growth, one has the same conclusion, except that not only $[0,\infty)$ but a whole sector $|\mathrm{Arg}\ z| \leq \theta$ must be excluded from the region where $G(z)$ is defined.

3. When $f = f_\chi$, $G(z)$ is a "twisted" (by χ) log gamma function. In fact, we have

Proposition. *Let* χ *be a nontrivial even Dirichlet character, and let* f *be defined by* (2.6). *Then the Stieltjes transform* G *of* f *is*

$$\int_0^\infty \frac{f_\chi(x)}{x+z}\,dx = \sum_{a=1}^{d} \chi(a) \log \Gamma\left(\frac{z+a}{d}\right). \qquad (2.10)$$

Proof. Let $G(z)$ be the left side of (2.10). Then

$$G(z) = \lim_{n\to\infty} \int_0^{nd} \frac{f_\chi(x)}{x+z}\,dx$$

$$= \lim_{n\to\infty} \sum_{j=0}^{dn-1} \left(\left(\log(z+j+1) - \log(z+j)\right) \sum_{a=1}^{j} \chi(a) \right)$$

$$= \lim_{n\to\infty} - \sum_{j=0}^{dn-1} \log(z+j)\ \chi(j)$$

$$= \lim_{n\to\infty} - \sum_{a=1}^{d} \chi(a) \sum_{j=0}^{n-1} \log(z+dj+a)$$

$$= \lim_{n\to\infty} - \sum_{a=1}^{d} \chi(a) \sum_{j=0}^{n-1} \log\left(\frac{z+a}{d}+j\right).$$

On the other hand, using the standard formula

$$\Gamma(z) = \lim_{n\to\infty} \frac{(n-1)!\,n^z}{z(z+1)\cdots(z+n-1)},$$

we see that the right side of (2.10) equals

$$\lim_{n\to\infty} \sum_{a=1}^{d} \chi(a) \left(\log(n-1)! + \frac{z+a}{d}\log n - \sum_{j=0}^{n-1} \log\left(\frac{z+a}{d}+j\right)\right),$$

and the first two terms vanish, because $\Sigma\,\chi(a) = \Sigma\, a\chi(a) = 0$ for χ nontrivial and even. This proves (2.10).

4. If χ is odd or the trivial character, then, in addition to the Stieltjes transform $G(z)$, the asymptotic series for the twisted log gamma function on the right in (2.10) also includes a principal term. For example, in the case of the trivial character, we have the "Stirling series" (see [97], p. 261)

$$\log\frac{\Gamma(z)}{\sqrt{2\pi}} = (z-\tfrac{1}{2})\log z - z - \int_0^\infty \frac{x-[x]-1/2}{x+z}\,dx.$$

Note that for the trivial character we take $f_{\chi_{triv}}(x) = \sum \frac{e^{2\pi inx}}{2\pi in} = -B_1(x) = [x]-x+\frac{1}{2}$. In other words, the Stieltjes transform of the first Bernoulli polynomial gives the error in Stirling's formula

$$n! \sim \sqrt{2\pi n}\,\frac{n^n}{e^n}.$$

5. As mentioned before, the classical Stieltjes transform can be defined more generally for a Stieltjes measure $\mu = dF(x)$ on the positive reals. For example, if χ is a nontrivial even Dirichlet character of conductor d, then the derivative of

$$\sum_{a=1}^{d} \chi(a) \log \Gamma\left(\frac{z+a}{d}\right) = -\lim_{n\to\infty} \sum_{j=0}^{dn-1} \chi(j) \log(z+j)$$

(see the proof of (2.10)) is

$$-\lim_{n\to\infty} \sum_{j=0}^{dn-1} \frac{\chi(j)}{z+j} = -\int_0^\infty \frac{df_\chi(x)}{x+z},$$

where f_χ is the function (2.6), i.e., df_χ has point mass $\chi(j)$ at j. Thus,

$$\frac{d}{dz} \sum_{a=1}^{d} \chi(a) \log \Gamma\left(\frac{z+a}{d}\right) = -\int_0^\infty \frac{df_\chi(x)}{x+z}. \tag{2.11}$$

This formula can also be obtained by differentiating (2.10) under the integral sign and then integrating by parts:

$$\frac{d}{dz} \sum_{a=1}^{d} \chi(a) \log \Gamma\left(\frac{z+a}{d}\right) = -\int_0^\infty \frac{f_\chi(x)}{(x+z)^2}\,dx$$

$$= \frac{f_\chi(x)}{x+z}\bigg|_0^\infty - \int_0^\infty \frac{df_\chi(x)}{x+z} = -\int_0^\infty \frac{df_\chi(x)}{x+z}.$$

The formula (2.11) is closely analogous to the formula for the derivative of the twisted p-adic log gamma function (see (8.6) in Chapter II). More precisely, define the p-adic log gamma function twisted by a nontrivial even character χ as follows:

$$G_{p,\chi}(z) \underset{\text{def}}{=} \sum_{a=1}^{d} \chi(a)\, G_p\left(\frac{z+a}{d}\right)$$

$$= \lim_{n\to\infty} p^{-n} \sum_{0<a<d,\ 0\le j<p^n} \chi(a)\left(\frac{z+a}{d}+j\right)\left(\ln_p\left(\frac{z+a}{d}+j\right)-1\right)$$

$$= \lim_{n\to\infty} \frac{1}{dp^n} \sum_{0<j<dp^n} \chi(j)\,(z+j)\left(\ln_p(z+j)-1\right).$$

Then $G_{p,\chi}$ can be expressed in terms of $G_{p,\xi}(z) = \lim_{n\to\infty} \frac{1}{dp^n} \sum_{0\le j<dp^n} \xi^j\,(z+j)\left(\ln_p(z+j)-1\right)$, where $\xi^d=1$, $\xi\neq 1$, as follows:

$$G_{p,\chi}(z) = \frac{g_\chi}{d} \sum_{b=1}^{d} \bar{\chi}(b)\; G_{p,\xi^b}(z),$$

where ξ is a fixed primitive d-th root of unity and $g_\chi = \sum_{a=1}^{d} \chi(a)\xi^a = d/g_{\bar{\chi}}$. So if we define a measure μ_χ on Z_p by

$$\mu_\chi = \frac{g_\chi}{d} \sum_{b=1}^{d} \bar{\chi}(b)\,\mu_{\xi^b}, \quad \text{where} \quad \mu_{\xi^b}(a+p^N Z_p) = \frac{\xi^{ab}}{1-\xi^{bp^N}}, \tag{2.12}$$

then we have

$$G_{p,\chi}(z) = -\int_{Z_p} \ln_p(x+z)\; d\mu_\chi(x);$$

$$\frac{d}{dz} \sum_{a=1}^{d} \chi(a)\; G_p\left(\frac{z+a}{d}\right) = -\int_{Z_p} \frac{d\mu_\chi(x)}{x+z},$$

which is the p-adic analog of (2.11).

Final remark. In the classical case an integer a prime to d has point mass

$$df_\chi(a) = \chi(a) = \frac{g_\chi}{d} \sum_{b=1}^{d} \bar{\chi}(b)\,\xi^{ab}. \tag{2.13}$$

Compare (2.13) to (2.12). As in our discussion of Leopoldt's formula for $L_p(1,\chi)$ in §II.5, we see that the p-adic construction is formally analogous to the classical case inside the open unit disc (in (2.12) note that $\mu_{\xi^b}(a+p^N Z_p) \longrightarrow \xi^{ab}$ as $N \longrightarrow \infty$), but the p-adic case only becomes arithmetically interesting when we extend to roots of unity, which are on the "boundary" of the unit disc.

In the remaining sections we shall give a systematic account of the p-adic Stieltjes transform, following Vishik [95]. First we introduce a type of p-adic integration (not to be confused with the type used in Chapter II) which is the tool used to construct the inverse Stieltjes transform in the p-adic case.

3. The Shnirelman integral and the p-adic Stieltjes transform

A p-adic analog of the line integral was introduced by Shnirelman in 1938 [88]. It can be used to prove p-adic analogs of the Cauchy integral theorem, the residue theorem, and the maximum modulus principle of complex analysis. The main applications of the Shnirelman integral are in transcendental number theory (see [1], [17]). Our interest in it will be to construct the inverse Stieltjes transform.

Definition. Let $f(x)$ be an Ω_p-valued function defined on all $x \in \Omega_p$ such that $|x-a|_p = r$, where $a \in \Omega_p$ and r is a positive real number. (We shall always assume that r is in $|\Omega_p|_p$, i.e., a rational power of p.) Let $\Gamma \in \Omega_p$ be such that $|\Gamma|_p = r$. Then the <u>Shnirelman integral</u> is defined as the following limit if it exists:

$$\int_{a,\Gamma} f(x)dx \underset{\text{def}}{=} \lim_{n\to\infty}{}' \frac{1}{n} \sum_{\xi^n=1} f(a+\xi\Gamma),$$

where the ' indicates that the limit is only over n not divisible by p.

Lemma 1. (1) *If $\int_{a,\Gamma} f(x)dx$ exists, then*

$$\left|\int_{a,\Gamma} f(x)dx\right|_p \le \max_{|x-a|_p=r} |f(x)|_p.$$

(2) $\int_{a,\Gamma}$ *commutes with limits of functions which are uniform limits on $\{x \mid |x-a|_p=r\}$.*

(3) *If $r_1 \le r \le r_2$ and $f(x)$ is given by a convergent Laurent*

series $\sum_{k=-\infty}^{\infty} c_k(x-a)^k$ in the annulus $r_1 \leq |x-a|_p \leq r_2$, then $\int_{a,\Gamma} f(x)dx$ exists and equals c_0. In particular, the integral does not depend on the choice of Γ with $|\Gamma|_p = r$ or even on r, as long as $r_1 \leq r \leq r_2$. More generally,

$$\int_{a,\Gamma} f(x)\ (x-a)^{-k}\ dx = c_k.$$

The proof of the lemma is easy. Part (3) uses the fact that if $k \neq 0$, then $\sum_{\xi^n=1} \xi^k = 0$ for $n > |k|$.

Lemma 2. For fixed $z \in \Omega_p$ and for $m > 0$

$$\int_{a,\Gamma} \frac{dx}{(x-z)^m} = \begin{cases} 0 & \text{if } |z-a|_p < r; \\ (a-z)^{-m} & \text{if } |z-a|_p > r. \end{cases}$$

To prove this, note that for $|x-a|_p = r$ we have the Laurent expansion

$$\frac{1}{(x-z)^m} = \begin{cases} \left(\sum_{k=0}^{\infty} (z-a)^k\ (x-a)^{-k-1}\right)^m & \text{if } |z-a|_p < r; \\ \left(\frac{1}{a-z} \sum_{k=0}^{\infty} (z-a)^{-k}\ (x-a)^k\right)^m & \text{if } |z-a|_p > r. \end{cases} \tag{3.1}$$

Then use part (3) of Lemma 1 (with $r_1 = r_2 = r$).

Lemma 3. (1) If $f(x)$ is a function on the closed disc of radius r with center a, i.e., $f\colon D_a(r) \longrightarrow \Omega_p$, and if $f(x) = \sum_{k=0}^{\infty} c_k(x-a)^k$ with $r^k|c_k|_p \longrightarrow 0$, define $\|f\|_r = \max_k r^k|c_k|_p$. Then $\max_{x \in D_a(r)} |f(x)|_p$ is attained when $|x-a|_p = r$ and equals $\|f\|_r$.

(2) Any Krasner analytic function $f\colon D_a(r) \longrightarrow \Omega_p$ (i.e., f is a uniform limit of rational functions with poles outside $D_a(r)$) is of the form in part (1), i.e., is given by a power series.

Proof. Making a linear change of variables, without loss of generality we may assume that $a = 0$ and $r = 1$. Multiplying f by a constant, we may also suppose that $\|f\|_1 = \max |c_k|_p = 1$. Clearly $|f(x)|_p = |\Sigma c_k x^k| \leq 1$ for $x \in D_0(1)$. Let $\bar{f}(x) = \Sigma\, \bar{c}_k x^k \in \bar{F}_p[x]$ be the reduction modulo M_{Ω_p}, the maximal ideal in Ω_p. If $x \in D_0(1)$ is any element whose reduction mod M_{Ω_p} is nonzero and is not a root of the polynomial $\bar{f}$, then $|x|_p = 1$ and $|f(x)|_p = 1$. This proves part (1).

It follows from part (1) that, if f_n is a sequence of rational functions approaching f uniformly, and if each f_n is represented by a power series on $D_a(r)$, then the sequence of power series approaches (coefficient by coefficient) a power series which represents f on $D_a(r)$. Thus, it suffices to consider the case when f is a rational function with poles outside $D_a(r)$. Again we make a change of variables so that $a = 0$ and $r = 1$. Using decomposition into partial fractions, we reduce to the case $f(x) = (x-b)^{-m-1}$, $|b|_p > 1$. But

$$(x-b)^{-m-1} = (-b)^{-m-1} \sum_{k=0}^{\infty} \binom{k+m}{m} \left(\frac{x}{b}\right)^k,$$

which converges on $D_0(1)$.

Lemma 4 (p-adic Cauchy integral formula). *If* f *is Krasner analytic in* $D_a(r)$, *and if* $|\Gamma|_p = r$, *then for fixed* $z \in \Omega_p$

$$\int_{a,\Gamma} \frac{f(x)\ (x-a)}{x-z}\, dx = \begin{cases} f(z) & \text{if } |z-a|_p < r; \\ 0 & \text{if } |z-a|_p > r. \end{cases} \tag{3.2}$$

In particular, this integral does not depend on the choice of a, Γ, *or* r *as long as* $|z-a|_p$ *remains either* $< r$ *or* $> r$. *More generally,*

$$\int_{a,\Gamma} \frac{f(x)(x-a)}{(x-z)^{m+1}}\, dx = \begin{cases} \frac{1}{m!} f^{(m)}(z) & \text{if } |z-a|_p < r; \\ 0 & \text{if } |z-a|_p > r. \end{cases} \tag{3.3}$$

Proof. By Lemma 3 and the linearity and continuity of both sides (part (2) of Lemma 1), we reduce to the case $f(x) = (x-a)^n$. Then write

$$\frac{1}{(x-z)^{m+1}} = \begin{cases} \sum_{k=m+1}^{\infty} \binom{k-1}{m} (z-a)^{k-m-1}(x-a)^{-k} & \text{if } |z-a|_p < r; \\ (-1)^{m+1}\sum_{k=0}^{\infty} \binom{k+m}{m} (z-a)^{-k-m-1}(x-a)^{k} & \text{if } |z-a|_p > r. \end{cases}$$

Now use part (3) of Lemma 1 to conclude (3.3).

Lemma 5 (p-adic residue theorem). _Let_ $f(x) = g(x)/h(x)$, _where_ $g(x)$ _is Krasner analytic in_ $D_a(r)$ _(i.e., by Lemma 3, a power series) and_ $h(x)$ _is a polynomial. Let_ $\{x_i\}$ _be the roots of_ h _in_ $D_a(r)$, _and suppose that all_ $|x_i - a|_p$ _are strictly_ $< r$. _Define_ $\mathrm{res}_{x_i} f$ _to be the coefficient of_ $(x - x_i)^{-1}$ _in the Laurent expansion of_ $f(x)$ _at_ x_i. _Then_

$$\int_{a,\Gamma} f(x)\,(x-a)\,dx = \sum \mathrm{res}_{x_i} f.$$

Proof. Using the partial fraction decomposition of $1/h(x)$, we reduce to the case $h(x) = (x - x_i)^{m+1}$. Then use (3.3) with $f(x)$ replaced by $g(x)$ and z replaced by x_i.

The next lemma will be stated and proved in the form we shall need it, although some of the assumptions can be eliminated (the D_i can have different radii, and $f(x)$ can approach a nonzero finite limit at infinity).

Lemma 6 (p-adic maximum modulus principle). _Let_ $f(x)$ _be a Krasner analytic function on_ $\Omega_p - \bigcup D_i$, _where_ $D_i = D_{a_i}(r^-)$ _are open discs of radius_ r. _Further suppose that_ $f(x) \longrightarrow 0$ _as_ $|x|_p \longrightarrow \infty$. _Then_ $|f(x)|_p$ _reaches its maximum on the boundary, i.e., if_ $|f(x)|_p \le M$ _for all_ x _with_ $|x - a_i|_p = r$ _for some_ i, _then_ $|f(x)|_p \le M$ _for all_ $x \in \Omega_p - \bigcup D_i$.

Proof. By the definition of Krasner analyticity, we immediately reduce to the case where $f(x)$ itself is a rational function with poles $b_j \in \bigcup D_i$. Let $z \in \Omega_p$ be such that $|z-a_i|_p > r$ for all i. We must show that $|f(z)|_p \le M$. Choose r_2 large enough so that $D_0(r_2) = D_z(r_2)$ contains z and all of the D_i, and so that $|f(x)|_p \le M$ for $|x|_p = r_2$. Let $|\Gamma_2|_p = r_2$. By part (1) of Lemma 1,

$$\left| \int_{z,\Gamma_2} f(x)dx \right|_p \le M. \tag{3.4}$$

On the other hand, by Lemma 5,

$$\begin{aligned} \int_{z,\Gamma_2} f(x)dx &= \sum \operatorname{res} \frac{f(x)}{x-z} \\ &= f(z) + \sum_j \operatorname{res}_{b_j} \frac{f(x)}{x-z}. \end{aligned} \tag{3.5}$$

Now let $|\Gamma|_p = r$. By Lemma 5, for each i

$$\sum_{b_j \in D_i} \operatorname{res}_{b_j} \frac{f(x)}{x-z} = \int_{a_i,\Gamma} f(x) \frac{x-a_i}{x-z} dx.$$

Since $|x-z|_p > |x-a_i|_p$ for $|x-a_i|_p = r$, it follows by part (1) of Lemma 1 that for each i

$$\left| \sum_{b_j \in D_i} \operatorname{res}_{b_j} \frac{f(x)}{x-z} \right|_p \le \max_{|x-a_i|_p = r} |f(x)|_p \le M.$$

Combining this with (3.4) and (3.5) gives $|f(z)|_p \le M$.

This concludes the basic lemmas relating to the Shnirelman integral.

Let $\sigma \subset \Omega_p$ be a compact subset, such as Z_p or Z_p^*. Let $\bar{\sigma} = \Omega_p - \sigma$ be its complement. For $z \in \bar{\sigma}$ let $\operatorname{dist}(z,\sigma)$ denote the minimum of $|z-x|_p$ as x ranges through σ.

Let $H_0(\bar{\sigma})$ denote the set of functions $\phi\colon \bar{\sigma} \longrightarrow \Omega_p$ which are Krasner analytic and vanish at infinity, i.e.,

(1) ϕ is a limit of rational functions whose poles are con-

tained in σ, the limit being uniform in any set $\overline{D}_\sigma(r) \underset{\text{def}}{=} \{z \in \Omega_p \mid$ $\text{dist}(z,\sigma) \geq r\}$.

(2) $\lim_{|z|_p \to \infty} \phi(z) = 0.$

Remark. Strictly speaking, to say that ϕ is Krasner analytic on $\bar{\sigma}$ a priori means only that for every $r > 0$ it is a uniform limit on $\overline{D}_\sigma(r)$ of rational functions ϕ_n with poles in $D_\sigma(r^-)$ $\underset{\text{def}}{=} \{z \in \Omega_p \mid \text{dist}(z,\sigma) < r\}$. But if, for example, $\phi_n(z) = 1/(z-b)$ with $|b-a|_p = r_1 < r$ for some $a \in \sigma$, then $\phi_n(z) = \sum_j \frac{(b-a)^j}{(z-a)^{j+1}}$ can be approximated uniformly on $\overline{D}_\sigma(r)$ by a rational function with pole $a \in \sigma$. Similarly if $\phi_n(z) = (z-b)^{-m}$; and any rational function ϕ_n can be reduced to this case by partial fractions. Thus, the poles of ϕ_n can be "shifted" to lie in σ.

Examples. 1. Since $\frac{d^2}{dz^2}\Big((z+j)(\ln_p(z+j) - 1)\Big) = \frac{1}{z+j}$, we have

$$\frac{d^2}{dz^2} G_p(z) = \lim_{n \to \infty} p^{-n} \sum_{0 \leq j < p^n} \frac{1}{z+j} \in H_0(\overline{Z}_p),$$

where G_p is Diamond's p-adic log gamma function, see (7.4) in Chapter II. (It is not hard to show that the construction $\lim p^{-n} \Sigma f(z+j)$ discussed in §II.7 commutes with differentiation when f is locally analytic.)

2. For any fixed $\xi \notin D_1(1^-)$, the first derivative of the twisted log gamma function $G_{p,\xi}$ (see (8.1) of Chapter II) is already Krasner analytic, since, by (8.6) of Chapter II,

$$\frac{d}{dz} G_{p,\xi}(z) = -\int_{Z_p} \frac{d\mu_\xi(x)}{x+z} = -\lim_{N \to \infty} \sum_{0 \leq a < p^N} \frac{1}{z+j} \frac{\xi^j}{1-\xi^{p^N}} \in H_0(\overline{Z}_p).$$

3. If μ is any measure on Z_p and $f(x) \in H_0(\overline{Z}_p)$, then it is not hard to check that $g(z) = \int_{Z_p} f(x-z)d\mu(x) \in H_0(\overline{Z}_p)$. In other words, $H_0(\overline{Z}_p)$ is stable under convolution with measures on Z_p. The function $\frac{d}{dz} G_{p,\xi}(-z)$ in Example 2 illustrates this.

For $r > 0$ the set $D_\sigma(r^-)$ is a finite (since σ is compact) disjoint union of open discs of radius r: $D_\sigma(r^-) = \bigcup D_{a_i}(r^-)$. For example, if $\sigma = Z_p$ and $r = p^{-n}$ there are p^{n+1} such discs.

Similarly, $D_\sigma(r) \underset{\text{def}}{=} \{z \in \Omega_p \mid |z-a|_p \le r$ for some $a \in \sigma\}$ is a finite disjoint union of $D_{a_i}(r)$, $a_i \in \sigma$.

Recall the definition $\overline{D}_\sigma(r) = \Omega_p - D_\sigma(r^-) = \{z \in \Omega_p \mid |z-a|_p \ge r$ for all $a \in \sigma\}$.

For $\phi \in H_0(\overline{\sigma})$ let $\|\phi\|_r \underset{\text{def}}{=} \max_{z \in \overline{D}_\sigma(r)} |\phi(z)|_p$. Obviously, $\|\phi\|_{r_1} \ge \|\phi\|_r$ if $r_1 < r$. By Lemma 6,

$$\|\phi\|_r = \max_{\text{dist}(z,\sigma)=r} |\phi(z)|_p.$$

We introduce a topology on $H_0(\overline{\sigma})$ by taking as a basis of open neighborhoods of zero

$$U(r,\varepsilon) = \{\phi \mid \|\phi\|_r < \varepsilon\}.$$

Note that $\|\phi\|_r$ is a continuous decreasing function of r. To see continuity, one easily reduces to the case when ϕ is a rational function with poles in σ, and then by partial fractions to the case when $\phi(x) = (x-a)^{-m}$, in which case it's obvious.

We next introduce a space of functions which will play a dual role to $H_0(\overline{\sigma})$ via a pairing defined using the Shnirelman integral.

Let

$$B_r = B_r(\sigma) \underset{\text{def}}{=} \{f\colon D_\sigma(r) \longrightarrow \Omega_p \mid f \text{ is Krasner analytic on each } D_{a_i}(r) \subset D_\sigma(r)\}.$$

By Lemma 3,

$$B_r = \{f \mid \text{on each } D_{a_i}(r),\ f \text{ is given by a convergent power series } \textstyle\sum_j c_{ij}(z-a_i)^j, \text{ i.e., } r^j|c_{ij}|_p \longrightarrow 0 \text{ as } j \longrightarrow \infty \text{ for each } i\}.$$

If $r_1 < r$, then restriction to $D_\sigma(r_1)$ gives an imbedding

$B_r \hookrightarrow B_{r_1}$. We denote $L(\sigma) = \bigcup_{r>0} B_r$. By definition (see the beginning of §II.7), $L(\sigma)$ is the set of <u>locally analytic functions</u> on σ.

For $f \in B_r$ we set

$$\|f\|_r \underset{\text{def}}{=} \max_{i,j} |c_{ij}|_p r^j,$$

which is finite by definition. By Lemma 3, $\|f\|_r = \max_{z \in D_\sigma(r)} |f(z)|_p$.

Note that the inclusion $B_r \hookrightarrow B_{r_1}$ for $r_1 < r$ is continuous with respect to $\| \ \|_r$ in B_r and $\| \ \|_{r_1}$ in B_{r_1}.

Let $L^*(\sigma)$ be the set of <u>continuous functionals</u> on $L(\sigma) = \bigcup B_r$, i.e., the set of linear maps μ (compatible with the restriction $B_r \hookrightarrow B_{r_1}$) such that for all r

$$\|\mu\|_r \underset{\text{def}}{=} \max_{0 \neq f \in B_r} |\mu(f)|_p / \|f\|_r$$

is finite. Note that $\|\mu\|_{r_1} \geq \|\mu\|_r$ if $r_1 < r$. We do <u>not</u> require that $\|\mu\|_r$ remain bounded as $r \longrightarrow 0$.

Key example. Let μ be a measure on σ, i.e., a bounded additive map from compact-open subsets U of σ to Ω_p. As in the case $\sigma = Z_p$ (see §II.2), the map

$$\mu: \ f \longmapsto \int_\sigma f \, d\mu \underset{\text{def}}{=} \lim_j \sum_i f_j(U_{ij}) \, \mu(U_{ij}) \tag{3.6}$$

(where f_j is a sequence of locally constant functions which approach f uniformly, and the U_{ij} are compact-open sets on which f_j is constant) is a well-defined functional on the continuous functions on σ, and <u>a fortiori</u> on $L(\sigma)$.

Lemma 7. $\mu \in L^*(\sigma)$ <u>comes from a measure on</u> σ <u>if and only if</u> $\|\mu\|_r$ <u>is bounded as</u> $r \longrightarrow 0$.

Proof. Using (3.6), it is easy to check that $\lim_{r \to 0} \|\mu\|_r = \max_U |\mu(U)|_p < \infty$ whenever μ is a measure. Conversely, suppose

$\mu \in L^*(\sigma)$ has $\|\mu\|_r \le M$ for all r. Define a function, also denoted μ, on compact-open subsets U of σ by

$$\mu(U) = \mu(\text{characteristic function of } U).$$

(Note that any locally constant function is in B_r for r small.) μ is obviously additive, and $|\mu(U)|_p \le \|\mu\|_r \cdot \|\text{char fn of } U\|_r = \|\mu\|_r \le M$ for all U. This proves the lemma.

Choose Γ with $|\Gamma|_p = r$, and define

$$f_{ij,r}(z) = \left(\frac{z - a_i}{\Gamma}\right)^j \text{ restricted to } D_{a_i}(r). \tag{3.7}$$

By Lemma 3, B_r is the set of all series $f = \Sigma\, c_{ij} f_{ij,r}$ with $c_{ij} \longrightarrow 0$ as $j \longrightarrow \infty$ for each of the (finitely many) i, and $\|f\|_r = \max_{i,j} |c_{ij}|_p$. For $\mu \in L^*(\sigma)$ clearly

$$\|\mu\|_r = \max_{i,j} |\mu(f_{ij,r})|_p.$$

It can then be shown that the weak topology in $L^*(\sigma)$, which has basis of neighborhoods of zero

$$V_{f,\varepsilon} \underset{\text{def}}{=} \{\mu \mid |\mu(f)|_p < \varepsilon\}, \tag{3.8}$$

is equivalent to the (a priori stronger) topology having basis of neighborhoods of zero

$$V(r,\varepsilon) \underset{\text{def}}{=} \{\mu \mid \|\mu\|_r < \varepsilon\}. \tag{3.9}$$

We shall prove this in the next section as a corollary of a general lemma on p-adic Banach spaces.

We shall often denote $\mu(f)$ by $(\mu(x), f(x))$.

Definition. For $\mu \in L^*(\sigma)$ the <u>Stieltjes transform</u> $S\mu: \bar{\sigma} \longrightarrow \Omega_p$ is the map

$$\phi: z \longmapsto \left(\mu(x), \frac{1}{z - x}\right). \tag{3.10}$$

We write $\phi = S\mu$.

Note that (3.10) makes sense, since for fixed $z \notin \sigma$, we have

$\frac{1}{z-x} \in B_r(\sigma)$ as soon as $r < \mathrm{dist}(z,\sigma)$.

Remark. If μ comes from a measure on σ (also denoted μ), then

$$S\mu(z) = \int_\sigma \frac{d\mu(x)}{z-x}.$$

This is slightly different from our earlier use of the term "Stieltjes transform" for the dlog gamma type functions $\frac{d}{dz}G$; namely, $\frac{d}{dz}G(z) = S\mu(-z)$.

Definition. For $\phi \in H_0(\bar{\sigma})$ the <u>Vishik transform</u> $V\phi$ of ϕ is the functional on $L(\sigma) = \cup B_r$ defined by

$$B_r \ni f \longmapsto \sum_i \int_{a_i,\Gamma} \phi(x)\, f(x)\, (x-a_i)dx, \tag{3.11}$$

where this integral is the Shnirelman integral defined at the beginning of the section.

Lemma 8. (3.11) <u>does not depend on the choice of centers</u> a_i <u>or the choice of</u> Γ <u>with</u> $|\Gamma|_p = r$, <u>and it is compatible with the inclusion</u> $B_r \hookrightarrow B_{r_1}$ <u>for</u> $r_1 < r$.

Proof. For fixed f, the right side depends continuously on ϕ (with respect to $\|\phi\|_r$), so we may reduce to the case when ϕ is a rational function with poles in σ. In that case, by Lemma 5, the right side of (3.11) is simply $\Sigma\, \mathrm{res}(\phi f)$, and the lemma follows.

Remark. A function $\phi \in H_0(\bar{\sigma})$ and a function $f \in B_r(\sigma)$ have an annulus around σ as a common domain of definition. The pairing $(\phi,f) = V\phi(f)$ can be thought of as a pairing which evaluates the sum of the residues of the product. For example, if σ is simply the point $\{0\}$, then $\phi(x) = \sum_{m<0} b_m x^m$, $f(x) = \sum_{n\geq 0} c_n x^n$, and

$$(\phi,f) = \text{coefficient of } \tfrac{1}{x} \text{ in } \phi(x)f(x) = \sum_{m+n=-1} b_m c_n.$$

Theorem (Vishik). V <u>and</u> S <u>are mutually inverse topological isomorphisms between</u> $H_0(\bar{\sigma})$ <u>and</u> $L^*(\sigma)$. <u>Under this isomorphism</u>

the subspace $M(\sigma) \subset L^*(\sigma)$ of measures on σ corresponds to the set of $\phi \in H_0(\bar{\sigma})$ such that $r\|\phi\|_r$ is bounded as $r \longrightarrow 0$.

Proof. Step 1. $S\mu \in H_0(\bar{\sigma})$, and S is continuous.

Notice that for fixed $z \in \bar{D}_\sigma(r)$ and for $r_1 < r$ and $|\Gamma|_p = r_1$, the image of $\frac{1}{z-x}$ in B_{r_1} is $\sum_{i,j} (z-a_i)^{-j-1} \Gamma^j f_{ij,r_1}(x)$ (where f_{ij,r_1} was defined in (3.7)). Then, since $|(\mu, f_{ij,r_1})|_p \le \|\mu\|_{r_1} \cdot \|f_{ij,r_1}\|_{r_1} = \|\mu\|_{r_1}$, it follows that

$$(\mu(x), \frac{1}{z-x}) = \lim_{n \to \infty} \sum_i \sum_{j<n} \frac{\Gamma^j}{(z-a_i)^{j+1}} (\mu, f_{ij,r_1})$$

is a uniform limit of rational functions with poles $a_i \in \sigma$ and value zero at infinity. At the same time we see continuity, since if $\mu \in V(r_1, \varepsilon)$, i.e., if $\|\mu\|_{r_1} < \varepsilon$, then

$$\begin{aligned} \|(\mu(x), \tfrac{1}{z-x})\|_r &= \max_{z \in \bar{D}_\sigma(r)} |(\mu(x), \tfrac{1}{z-x})|_p \le \max_{i,j} \frac{r_1^j}{r^{j+1}} \|u\|_{r_1} \\ &< \frac{\varepsilon}{r}, \end{aligned} \tag{3.12}$$

in other words, $S\mu \in U(r, \varepsilon/r)$.

Step 2. For $\phi \in H_0(\bar{\sigma})$, the functional $V\phi$ is continuous, i.e., V maps $H_0(\bar{\sigma})$ to $L^*(\sigma)$.

Let $f \in B_r(\sigma)$. Then

$$\begin{aligned} |(V\phi, f)|_p &= |\sum_i \int_{a_i,\Gamma} \phi(x)\, f(x)\, (x-a_i) dx|_p \\ &\le \max_i \max_{|x-a_i|_p = r} |\phi(x) f(x) (x-a_i)|_p \end{aligned}$$

by part (1) of Lemma 1. But this is at most

$$r \max_{\mathrm{dist}(x,\sigma)=r} |\phi(x)|_p \max_{\mathrm{dist}(x,\sigma)=r} |f(x)|_p = r\|\phi\|_r \|f\|_r .$$

Step 3. $V\colon H_0(\bar{\sigma}) \longrightarrow L^*(\sigma)$ is continuous.

If $\phi \in U(r, \varepsilon)$, then we just saw that $|(V\phi, f)|_p < r\varepsilon \|f\|_r$

for $f \in B_r$. Thus $\|V\phi\|_r < r\varepsilon$, i.e., $V\phi \in V(r, r\varepsilon)$.

Step 4. $VS =$ identity.

To see this, let $\mu \in L^*(\sigma)$, $f \in B_r(\sigma)$, and denote $\phi = S\mu$. We must show that $(V\phi, f) = (\mu, f)$.

We have

$$(V\phi, f) = \sum_i \int_{a_i, \Gamma} (z-a_i) f(z) \phi(z) dz$$

$$= \sum_i \int_{a_i, \Gamma} (z-a_i) f(z) \left(\mu(x), \frac{1}{z-x}\right) dz.$$

Since μ is linear and continuous, it commutes with the Shnirelman integral, and we have

$$(V\phi, f) = \left(\mu(x), \sum_i \int_{a_i, \Gamma} f(z) \frac{z-a_i}{z-x} dz\right). \tag{3.13}$$

Without loss of generality we may suppose that $r \notin \{|a-b|_p \mid a, b \in \sigma\}$, in which case r_1 can be chosen less than r but close enough so that $D_\sigma(r_1)$ is obtained from $D_\sigma(r)$ by merely shrinking each of the discs $D_{a_i}(r)$ (i.e., no elements of σ are lost, so no new discs have to be added to cover σ). Now for $x \in D_{a_i}(r_1)$ the integral on the right in (3.13) is equal to $f(x)$, by Lemma 4. Thus, the restriction of $\sum_i \int_{a_i, \Gamma} f(z) \frac{z-a_i}{z-x} dz$ to B_{r_1} is the same as the restriction of $f(x)$. Hence, $(V\phi, f) = (\mu, f)$.

Step 5. $SV =$ identity.

Let $\phi \in H_0(\bar{\sigma})$.

We first suppose that z is large, say $|z|_p > r = |\Gamma|_p$, where r is taken large enough so that $\sigma \subset D_0(r) = D_\sigma(r)$. Let a be any point in σ. It is easy to see that $\phi(x) = \sum_{j=0}^{\infty} c_j (x-a)^{-j-1}$ for $x \in \overline{D}_\sigma(r)$ (as in the proof of Lemma 3).

Then

$$SV\phi(z) = (V\phi(x), \frac{1}{z-x}) = \int_{a,\Gamma} (x-a)\phi(x)\frac{1}{z-x}dx$$

$$= \int_{0,\Gamma} x\,\phi(x)\frac{1}{z-x}\,dx$$

$$= \int_{0,1/\Gamma} \frac{1}{x}\phi(\frac{1}{x})\frac{1}{z-1/x}\,dx$$

by the definition of the Shnirelman integral. Thus,

$$SV\phi(z) = \frac{1}{z}\int_{0,1/\Gamma} \phi(\frac{1}{x})\frac{1}{x-1/z}\,dx$$

$$= \frac{1}{z}\,\mathrm{res}_{1/z}\,\frac{1}{x}\frac{\phi(1/x)}{x-1/z} \qquad \text{by Lemma 5}$$

$$= \phi(z).$$

(Alternately, we could expand $\phi(x) = \Sigma\, c_j x^{-j-1}$ and compute

$$SV\phi(z) = \sum c_j \int_{0,\Gamma} x^{-j}\frac{1}{z-x}dx = \sum c_j \int_{0,\Gamma} \sum_{k=0}^{\infty} \frac{x^{k-j}}{z^{k+1}}dx$$

$$= \sum c_j/z^{j+1} = \phi(z).\ )$$

By Step 1, we know that $SV\phi(z) \in H_0(\bar{\sigma})$. Since $SV\phi(z)$ and $\phi(z)$ are both Krasner analytic on $\bar{\sigma}$ and agree on $\bar{D}_0(r)$, they must agree everywhere.

Step 6. <u>If</u> $\mu \in L^*(\sigma)$ <u>and</u> $\phi = S\mu$, <u>then</u> $\|\mu\|_r = r\|\phi\|_r$.

In Step 1 we saw that for any $r_2 > r$

$$\|\phi\|_{r_2} \le \frac{1}{r_2}\,\|\mu\|_r.$$

(We have replaced r and r_1 by r_2 and r, respectively, in (3.12).) Letting $r_2 \longrightarrow r$ and using the fact that $\|\phi\|_{r_2}$ is continuous in r_2, we obtain

$$r\,\|\phi\|_r \le \|\mu\|_r.$$

On the other hand, in Step 2 we saw that $|(\mu,f)|_p \le r\|\phi\|_r \cdot \|f\|_r$ for all $f \in B_r$, and hence $\|\mu\|_r \le r\,\|\phi\|_r$.

Step 7. $S(M(\sigma)) = \{\phi \in H_0(\bar{\sigma}) \mid \; r\|\phi\|_r \text{ is bounded}\}$.

This follows immediately from Step 6 and Lemma 7.

The proof of the theorem is now complete.

Remark. Amice and Vélu [5] and Vishik [94] have studied so-called "h-admissible measures" on σ. These are elements $\mu \in L^*(\sigma)$ which instead of boundedness are required to satisfy the weaker condition

$$r^{h-j}\,|\mu(f_{ij,r})|_p \longrightarrow 0 \quad \text{as} \quad r \longrightarrow 0 \quad \text{for all } i, j$$

($f_{ij,r}$ is the function (3.7)). For example, when $\sigma = Z_p$, $j = 0$, and $r = p^{-N}$, so that $f_{ij,r}$ is the characteristic function of some $a + p^N Z_p$, this condition says that $|\mu(a + p^N Z_p)|_p$ grows slower than p^{hN}. It is not hard to show that h-admissible measures μ correspond to functions $\phi \in H_0(\bar{\sigma})$ for which $r^{h+1}\|\phi\|_r$ approaches zero as $r \longrightarrow 0$.

Even the broader class of h-admissible measures are only a small part of $L^*(\sigma)$. For example, when $\sigma = \{0\}$, then $M(\sigma)$ is simply the constants, which correspond to elements $\phi \in H_0(\bar{\sigma})$ of the form $\phi(z) = \frac{\text{const}}{z}$. The h-admissible measures on $\{0\}$ correspond to the polynomials of degree at most h in $1/z$ (with no constant term); while $L^*(\sigma)$ corresponds to all series $\Sigma\, c_j z^{-j}$ for which $r^{-j}|c_j|_p \longrightarrow 0$ as $j \longrightarrow \infty$ for every r.

4. p-adic spectral theorem

We start by discussing p-adic Banach spaces. For a more complete account, see [82]. In the process we fill in a technical gap in the last section, namely, we prove that in $L^*(\sigma)$ the topology determined by

$$V_{f,\varepsilon} = \{\mu \mid \; |\mu(f)|_p < \varepsilon\} \tag{4.1}$$

is equivalent to the topology determined by

$$V(r,\varepsilon) = \{\mu \mid \; \|\mu\|_r < \varepsilon\}. \tag{4.2}$$

Let K be a field which is complete under a non-archimedean norm $|\ |_p$ (in practice, K will be a subfield of Ω_p).

Definition. A <u>Banach space</u> over K is a vector space B supplied with a norm $\|\ \|$ from B to the nonnegative real numbers such that for all $x, y \in B$ and $a \in K$: (1) $\|x\| = 0$ if and only if $x = 0$; (2) $\|x+y\| \leq \max(\|x\|, \|y\|)$; (3) $\|ax\| = |a|_p\|x\|$; (4) B is complete with respect to $\|\ \|$.

We shall also assume that $\|B\| = |K|_p$, i.e., for every $x \neq 0$ in B there exists $a \in K$ such that $\|ax\| = 1$.

By $\mathrm{Hom}(B_1,B_2)$ we mean the vector space of K-linear continuous maps from B_1 to B_2; $\mathrm{Hom}(B_1,B_2)$ is clearly a Banach space under the usual operator norm. We denote $\mathrm{End}(B) = \mathrm{Hom}(B,B)$.

If B_0 is a Banach space over $K_0 \subset K$, by $B_K = B_0 \hat{\otimes} K$ we mean the <u>completed tensor product</u>, i.e., the completion of the vector space $B_0 \otimes_{K_0} K$.

Example. If $B = \{f = \Sigma\, c_j x^j \in Q_p[[x]] \mid |c_j|_p \longrightarrow 0\}$, with $\|f\| = \max_j |c_j|_p$, then $B_{\Omega_p} = \{f = \Sigma\, c_j x^j \in \Omega_p[[x]] \mid |c_j|_p \longrightarrow 0\}$.

In practice, most interesting Ω_p-Banach spaces B are really defined over a finite extension K of Q_p, in the sense that $B = B_0 \hat{\otimes} \Omega_p$ for some K-Banach space B_0.

Canonical example. Let J be any indexing set, and let $K(J)$ denote the set of all sequences $c = \{c_j\}_{j \in J}$ such that for every ε only finitely many $|c_j|_p$ are $> \varepsilon$. Let $\|c\| = \max_j |c_j|_p$.

Note that $K(J) \hat{\otimes} \Omega_p = \Omega_p(J)$.

Proposition. <u>Let</u> K <u>be a discrete valuation ring (for example, a finite extension of</u> Q_p, <u>or the unramified closure of</u> Q_p<u>). Then any Banach space</u> B <u>over</u> K <u>is of the form (i.e., isomorphic to)</u> $K(J)$ <u>for some</u> J.

Proof. Let $0 = 0_K = \{a \in K \mid |a|_p \le 1\}$, $M = M_K = \{a \in K \mid |a|_p < 1\}$ $= \pi 0$, $k = 0/M$. Let $E = \{x \in B \mid \|x\| \le 1\}$, $\overline{E} = E/\pi E$. Let $\{e_j\}_{j \in J}$ be elements of E whose reductions mod πE form a basis for the k-vector space $\overline{E}$. We claim that B is isomorphic to $K(J)$. Given $x \in B$, find $a \in K$ such that $\|ax\| \le 1$. Then for some $\{c_{1j}\}_{j \in J}$ with $|c_{1j}|_p \le 1$ and only finitely many c_{1j} nonzero we have: $ax - \Sigma c_{1j} e_j \in \pi E$. Repeating this process for $\frac{1}{\pi}(ax - \Sigma c_{1j} e_j)$, we successively find $ax = \sum_j (\sum_i \pi^i c_{ij}) e_j$. Let $c_j = \frac{1}{a} \sum_i \pi^i c_{ij}$, and let x correspond to $\{c_j\} \in K(J)$. Conversely, let every $\{c_j\} \in K(J)$ correspond to $\Sigma c_j e_j$. It is easy to see that this correspondence gives an isomorphism $B \simeq K(J)$.

Such a set $\{e_j\} \subset B$ is called a "Banach basis" for B.

Example. The space $B_r(\sigma)$ in the last section has Banach basis $f_{ij,r}$ (see (3.7)).

Corollary. If an Ω_p-Banach space B is defined over a finite extension of Q_p, then B is isomorphic to $\Omega_p(J)$ for some J.

Definition. The dual space B^* of a K-Banach space B is $\mathrm{Hom}(B,K)$, which is a Banach space with the usual operator norm.

Lemma 1. If $B \simeq K(J)$, then B^* is isomorphic to the Banach space of bounded sequences $b = \{b_j\}_{j \in J}$ with $\|b\| \underset{\text{def}}{=} \max |b_j|_p$.

In fact, if $\{e_j\}$ is a Banach basis, let $\{b_j\}$ be the map $\Sigma c_j e_j \longmapsto \Sigma b_j c_j$. It is routine to check that this identifies B^* with the space of bounded sequences.

Definition. Let B be a K-Banach space. A sequence $x_1, x_2, x_3, \ldots$ is said to be weakly convergent to x if $h(x_i) \longrightarrow h(x)$ for all $h \in B^*$.

Lemma 2. Suppose $B \simeq K(J)$. If $x_i \longrightarrow x$ weakly, then

$\|x_i - x\| \longrightarrow 0.$

Proof. Since any countable set of elements of B is in the Banach subspace generated by a countable subset of our Banach basis for B, without loss of generality we may assume that J is the positive integers. Replacing x_i by $x_i - x$, without loss of generality we may suppose that $x = 0$.

Suppose $\|x_i\|$ does not approach zero. By passing to a subsequence, we may suppose that $\|x_i\| > \varepsilon$ for all i. We identify B with $K(J)$, and let $x_i = \{a_{ij}\} \in K(J)$. Then $\|x_i\| = \max_j |a_{ij}|_p$. Let α_i denote the first j for which $|a_{ij}|_p = \|x_i\|$, and let β_i denote the last j for which $|a_{ij}|_p = \|x_i\|$.

Case 1. α_i is bounded.

Then there exists some j_0 such that $\alpha_i = j_0$ for infinitely many i. Let $h \in B^*$ be the j_0-th coordinate map. Then for infinitely many i we have

$$|h(x_i)|_p = |a_{ij_0}|_p = |a_{i\alpha_i}|_p = \|x_i\| > \varepsilon,$$

a contradiction.

Case 2. α_i is unbounded.

Let

$j_0 = \alpha_1$

$j_1 = \alpha_{i_1}$, where i_1 is chosen so that $\alpha_{i_1} > \beta_1$

$j_2 = \alpha_{i_2}$, where i_2 is chosen so that $\alpha_{i_2} > \beta_{i_1}$

$\vdots$

$j_n = \alpha_{i_n}$, where i_n is chosen so that $\alpha_{i_n} > \beta_{i_{n-1}}$.

Let $h \in B^*$ be the sum of the j_n-th coordinate maps, i.e., $h(\{a_j\}) = \sum_n a_{j_n}$. Then for all m

$$|h(x_{i_m})|_p = |\sum_n a_{i_m j_n}|_p = |a_{i_m \alpha_{i_m}}|_p = \|x_{i_m}\| > \varepsilon,$$

and again $h(x_i)$ fails to approach 0. This concludes the proof.

Corollary 1. _If_ $B \simeq K(J)$ _and_ $\{x_i\}$, $x_i \in B$, _has the property that_ $h(x_i)$ _approaches a finite limit for all_ $h \in B^*$, _then_ $\{x_i\}$ _converges in the norm to some_ x.

Proof. Let $y_i = x_i - x_{i+1}$. By Lemma 2, $\|y_i\| \longrightarrow 0$. But then $\{x_i\}$ is a Cauchy sequence (since $\|x_M - x_N\| \le \max_{M \le i < N} \|x_i - x_{i+1}\|$), and the corollary follows from the completeness of B.

Corollary 2. _The topologies on_ $L^*(\sigma)$ _determined by_ (4.1) _and_ (4.2) _are equivalent._

Proof. Since the $V(r,\varepsilon)$-topology is trivially stronger than the $V_{f,\varepsilon}$-topology, it suffices to show that a sequence μ_k which converges to zero in the $V_{f,\varepsilon}$-topology must converge to zero in the $V(r,\varepsilon)$-topology. Suppose that for every $f \in L(\sigma) = \bigcup B_r$ we have $\mu_k(f) \longrightarrow 0$. We must show that for every r, $\|\mu_k\|_r \longrightarrow 0$.

Without loss of generality we may suppose that $r \notin \{|a-b|_p \mid a, b \in \sigma\}$.

For any r, $L^*(\sigma)$ maps to the dual $B_r^* \simeq \Omega_p(J)^*$, where J indexes the $f_{ij,r}$. Namely, $\mu \longmapsto \{\mu(f_{ij,r})\}_{i,j}$, and it is easy to see that the norm in B_r^* corresponds to $\|\mu\|_r$. Note that the image of μ has coordinates which approach zero as $j \longrightarrow \infty$ for each i. This is because, if we choose $r_1 < r$ but large enough so that $\bigcup_i D_{a_i}(r_1)$ still contains σ (this can be done because σ is compact, and no $b \in \sigma$ has $|b-a_i|_p = r$), then for all i, j

$$|\mu(f_{ij,r})|_p = |\mu(f_{ij,r}|_{D_\sigma(r_1)})|_p = |\mu(\left(\frac{\Gamma_1}{\Gamma}\right)^j f_{ij,r_1})|_p \qquad \text{where } |\Gamma_1|_p = r_1$$

$$= \left(\frac{r_1}{r}\right)^j |\mu(f_{ij,r_1})|_p \le \left(\frac{r_1}{r}\right)^j \|\mu\|_{r_1},$$

which approaches zero as $j \longrightarrow \infty$. In other words, the compatibility requirement with $B_r \hookrightarrow B_{r_1}$ forces μ to be a very special element of B_r^*.

Now the subspace of B_r^* consisting of elements whose coordinates approach zero is of course isomorphic to $\Omega_p(J)$. By Lemma 2, to show that $\|\mu_k\|_r \longrightarrow 0$ it suffices to show that $g(\mu_k) \longrightarrow 0$ for all $g \in \Omega_p(J)^*$. But if $g = \{g_{ij}\}$ under the isomorphism in Lemma 1, then

$$g(\mu_k) = \sum_{i,j} g_{ij}\ \mu_k(f_{ij,r}) = \sum_{i,j} g_{ij}\left(\frac{\Gamma_1}{\Gamma}\right)^j\ \mu_k(f_{ij,r_1}) = \mu_k(f),$$

where $f = \sum g_{ij}\,\Gamma_1^j\,\Gamma^{-j}\ f_{ij,r_1} \in B_{r_1}$. And $\mu_k(f) \longrightarrow 0$ by assumption.

We now discuss operators $A \in \mathrm{End}(B)$.

If $B = K(J)$ with Banach basis $\{e_j\}_{j \in J}$, then A corresponds to a matrix $\{a_{ij}\}_{i,j \in J}$ in the usual way:

$$Ae_j = \sum a_{ij} e_i.$$

It is easy to check that this is a norm-preserving isomorphism between $\mathrm{End}(B)$ and the Banach space of matrices $\{a_{ij}\}$ having finite $\|\{a_{ij}\}\| \underset{\text{def}}{=} \max_{i,j} |a_{ij}|_p$ and having the property that for each j, $a_{ij} \longrightarrow 0$ as $i \longrightarrow \infty$. In other words, when $B = K(J)$, A can be thought of as a matrix whose columns are in B and whose rows are in B^*.

An operator A is said to be "completely continuous" if it can be approximated by operators having finite-dimensional image. In terms of matrices, this means that $a_{ij} \longrightarrow 0$ as $i \longrightarrow \infty$ <u>uniformly in</u> j; in other words, the norm of the i-th row approaches zero. Such operators occur in Dwork's theory (e.g. [25]), and in [82] Serre gives a Riesz and Fredholm theory for them.

However, many simple operators are not completely continuous: the identity operator, for example, or the operator $\left(x\frac{d}{dx}\right)^n$ on $\{\Sigma c_i x^i \mid c_i \longrightarrow 0\}$ (which has diagonal matrix $a_{ii} = i^n$).

For simplicity, we shall assume our Banach spaces are of the form $\Omega_p(J)$. As mentioned before, all Ω_p-Banach spaces which are defined over a discrete valuation subfield of Ω_p are of this form.

Definition. For $A \in \mathrm{End}(B)$ let $\sigma_A = \{\lambda \in \Omega_p \mid A-\lambda$ does not have an inverse in $\mathrm{End}(B)\}$ denote the <u>spectrum</u> of A.

Definition. An operator $A \in \mathrm{End}(B)$ is called <u>analytic</u> if the "resolvent" operator $R_A(z) = (z-A)^{-1}$ is Krasner analytic in the complement of σ_A, in the sense that for all $x \in B$ and $h \in B^*$, $h(R_A(z)x)$ as a function of z lies in $H_0(\bar{\sigma}_A)$. If $B = \Omega_p(J)$, then in terms of matrices this is equivalent to the condition that each matrix entry in $R_A(z)$ be a Krasner analytic function of z on $\bar{\sigma}_A$ (and vanish at infinity).

Vishik's spectral theory applies to <u>analytic</u> operators A whose spectrum σ_A is a <u>compact</u> subset of Ω_p.

Example. $x\frac{d}{dx}$ acting on $B = \{\Sigma\, c_i x^i \mid c_i \longrightarrow 0\}$ has spectrum Z_p, and its resolvent is Krasner analytic on $\bar{Z}_p$.

It is possible for A to have a compact spectrum but not satisfy the analyticity condition. Here is an example of Vishik where the spectrum is <u>empty</u>. (Since the only Krasner analytic functions on all of Ω_p, by Lemma 3 of §3, are everywhere convergent power series, and since only the zero power series has value 0 at infinity, it follows that in the case of an empty spectrum $R_A(z)$ has no chance of having matrix entries in $H_0(\bar{\sigma}_A)$.)

Example. Let B be the set of $\{a_i\}_{i\in Z}$ such that $\|\{a_i\}\| \underset{\mathrm{def}}{=} \max_{i\in Z} |a_i|_p$ is finite and $a_i \longrightarrow 0$ as $i \longrightarrow -\infty$. Let A be the shift operator $A(\{a_i\}) = \{b_i\}$ where $b_i = a_{i+1}$.

Claim. For all $z \in \Omega_p$, $(z-A)$ has a continuous inverse f_z.

Proof of claim. We want to find $f_z: \{b_j\} \longmapsto \{a_i\}$ such that $a_i = f_{z,i}(\{b_j\})$ satisfies $za_i - a_{i+1} = b_i$.

Case (1). $|z|_p \leq 1$. Set $a_i = -b_{i-1} - zb_{i-2} - z^2 b_{i-3} - \cdots$, which converges because $b_j \longrightarrow 0$ as $j \longrightarrow -\infty$. Clearly $\{a_i\} \in B$; $za_i - a_{i+1} = b_i$; and this map f_z is continuous.

Case (2). $|z|_p > 1$. Set $a_i = b_i z^{-1} + b_{i+1} z^{-2} + \cdots$. Again $\{a_i\} \in B$; $za_i - a_{i+1} = b_i$; and the map is continuous.

Let B be a Banach space of the form $\Omega_p(J)$. Let $F(x)$ be an analytic operator-valued function on the complement of a compact set σ, i.e., for all $y \in B$ and $h \in B^*$ the Ω_p-valued function

$$F_{h,y}(x) \underset{\text{def}}{=} h(F(x)y)$$

belongs to $H_0(\bar{\sigma})$. Let $a \in \sigma$, $|\Gamma| = r$, and suppose that there are no $b \in \sigma$ such that $|b-a|_p = r$.

Definition. Let $S_n = \frac{1}{n} \sum_{\xi^n=1} F(a+\xi\Gamma)$. Then

$$\int_{a,\Gamma} F(x)\,dx \underset{\text{def}}{=} \lim_{n\to\infty,\ p\nmid n} S_n. \tag{4.3}$$

Lemma 3. The limit (4.3) converges in the operator norm to a continuous operator.

Proof. Let $y \in B$. For all $h \in B^*$, since $F_{h,y} \in H_0(\bar{\sigma})$, it follows that the ordinary Shnirelman integral $\int_{a,\Gamma} F_{h,y}(x)\,dx$ exists. That is, $h(S_n y)$ approaches a finite limit for all h. By Corollary 1 to Lemma 2, $S_n y$ converges in the norm. By the uniform boundedness principle, S_n converges to a continuous operator.

Note that from the proof of Lemma 3 it follows that

$$h\left(\int_{a,\Gamma} F(x)\,dx\ y\right) = \int_{a,\Gamma} F_{h,y}(x)\,dx. \tag{4.4}$$

Spectral theorem (Vishik). Let $B \simeq \Omega_p(J)$, and let $A \in \mathrm{End}(B)$ be analytic with compact spectrum σ_A. Then the operator-valued distribution

$$\mu_A \underset{\text{def}}{=} VR_A$$

where V is the Vishik transform in §3, gives a continuous homomorphism from the algebra $L(\sigma_A)$ to the algebra $\mathrm{End}(B)$. For $f \in B_r(\sigma_A)$, the operator $\mu_A(f)$ is defined as

$$\sum_i \int_{a_i,\Gamma} f(x)\,(x-a_i)\,(x-A)^{-1}\,dx,$$

where $D_\sigma(r) = \bigcup D_{a_i}(r)$ is a covering of σ by discs of radius r. In addition, the following inversion formula holds:

$$R_A(z) = \left(\mu_A(x), \frac{1}{z-x}\right) \tag{4.5}$$

Corollary 1. For all $j \geq 0$, $A^j = (\mu_A(x), x^j)$.

Proof of corollary. For any fixed z with $|z|_p > \max_{x\in\sigma_A} |x|_p$ and $|z|_p > \|A\|$, we can write $\frac{1}{z-x} = \sum_{j=0}^{\infty} z^{-j-1}x^j$ in (4.5). Since $|z|_p > \|A\|$, we also have $R_A(z) = \sum_{j=0}^{\infty} z^{-j-1}A^j$. By the continuity of μ_A: $L(\sigma_A) \longrightarrow \mathrm{End}(B)$, this gives us $\sum z^{-j-1}A^j = \sum z^{-j-1}(\mu_A(x), x^j)$. Since this holds for all large z, the coefficients can be equated, and the corollary is proved.

Remark. For $j=1$, if we write (μ_A, f) using the $\int$ notation, we obtain the usual form for a spectral theorem:

$$A = \int_{\sigma_A} x\, d\mu_A(x).$$

Proof of the spectral theorem. First of all, it is easy to see that $\mu_A(f)$ is a bounded operator, and that μ_A: $L(\sigma_A) \longrightarrow \mathrm{End}(B)$ is continuous. The key assertion is multiplicativity:

$$\mu_A(f_1f_2) = \mu_A(f_1)\,\mu_A(f_2) \quad \text{for} \quad f_1, f_2 \in L(\sigma_A).$$

Suppose that $f_1, f_2 \in B_r$. Let $r > r' > r_1$, $|\Gamma'|_p = r'$, $|\Gamma_1|_p = r_1$. We can choose r' so that $r' \notin \{|a-b|_p \mid a, b \in \sigma_A\}$, in which case r_1 can be chosen so that $D_{\sigma_A}(r_1)$ is obtained from $D_{\sigma_A}(r') = \bigcup D_{a_i}(r')$ by merely shrinking each disc. Thus, $D_{\sigma_A}(r_1)$

$= \bigcup D_{a_i}(r_1)$. Now

$$\mu_A(f_1)\,\mu_A(f_2) = \sum_i \sum_j \int_{a_j,\Gamma'} \int_{a_i,\Gamma_1} \frac{(x'-a_j)(x-a_i)f_2(x')f_1(x)}{(x'-A)\,(x-A)} dx dx'.$$

Since $\frac{1}{(x'-A)(x-A)} = \frac{1}{x-x'}\left((x'-A)^{-1} - (x-A)^{-1}\right)$, this equals

$$\sum_i \sum_j \int_{a_j,\Gamma'} (x'-a_j)f_2(x')(x'-A)^{-1} \int_{a_i,\Gamma_1} \frac{(x-a_i)f_1(x)}{x-x'} dx\, dx'$$

$$+ \sum_i \sum_j \int_{a_i,\Gamma_1} (x-a_i)f_1(x)(x-A)^{-1} \int_{a_j,\Gamma'} \frac{(x'-a_j)f_2(x')}{x'-x} dx'\, dx.$$

But by Lemma 4 of the last section, the inner integral in the first sum is zero, and the inner integral in the second sum is zero for $j \neq i$ and is $f_2(x)$ for $j = i$. Thus,

$$\mu_A(f_1)\,\mu_A(f_2) = \sum_i \int_{a_i,\Gamma_1} (x-a_i)f_1(x)f_2(x)\,(x-A)^{-1} dx = \mu_A(f_1 f_2).$$

Finally, to prove the inversion formula, for any $y \in B$ and $h \in B^*$ we have

$$h(R_A(z)\,y) = R_{A,h,y}(z) = SVR_{A,h,y}(z) \quad \text{by the theorem in §3}$$

$$= \left(VR_{A,h,y}(x), \frac{1}{z-x}\right) = h\left(\left(\mu_A(x), \frac{1}{z-x}\right) y\right).$$

Thus, (4.5) is an immediate consequence of the theorem on inverting the p-adic Stieltjes transform. This completes the proof of the theorem.

Corollary 2. Under the conditions of the theorem, the following two conditions are equivalent: (1) $\operatorname{dist}(z,\sigma)\,\|R_A(z)\|$ is bounded as $z \longrightarrow \sigma$; (2) μ_A is a projection-valued measure, i.e., a bounded homomorphism from the Boolean algebra of compact-open subsets of σ_A to the algebra End(B). In this case

$$\max_{z \in \bar{\sigma}_A} \left(\operatorname{dist}(z,\sigma)\,\|R_A(z)\|\right) = \max_U \|\mu_A(U)\|.$$

The proof is exactly like Steps 6 and 7 in the proof of the theorem in §3.

Corollary 3. *Let* B *and* $A \in End(B)$ *be as in the theorem. For all* $r > 0$ *the resolvent* $R_A(z)$ *can be uniformly approximated on* $\overline{D}_{\sigma_A}(r)$ *by rational functions* $\sum_i \sum_{j=1}^{N} A_{ij} (z-a_i)^{-j}$ *with operator coefficients and with poles in* σ_A.

The proof is just like Step 1 in the proof of the theorem in §3.

Remarks. 1. One could alternately take Corollary 3 as the *definition* of an analytic operator, in which case the spectral theorem would hold for an arbitrary Banach space. However, in practice the "weaker" definition is often easier to check than the strong condition in the corollary.

2. The operators in Corollary 2 are the closest p-adic analogs of normal operators or operators of scalar type [24] in a Hilbert space.

3. It is not hard to prove that operators for which $dist(z, \sigma_A)^{h+1} \|R_A(z)\| \longrightarrow 0$ as $z \longrightarrow \sigma$ correspond to "h-admissible" μ_A (see the remark at the end of the last section).

4. Vishik has also proved a generalization to functions of (the spectra of) several analytic operators. Namely, let $A_1, \ldots, A_n \in End(B)$ be commuting analytic operators with compact spectra $\sigma_{A_1}, \ldots, \sigma_{A_n}$. Let $\sigma = \sigma_{A_1} \times \cdots \times \sigma_{A_n} \subset \Omega_p^n$, and use the completed tensor product to define $B_r(\sigma)$ and $L(\sigma)$: $B_r(\sigma) = B_r(\sigma_{A_1}) \hat{\otimes} \cdots \hat{\otimes} B_r(\sigma_{A_n})$, $L(\sigma) = L(\sigma_{A_1}) \hat{\otimes} \cdots \hat{\otimes} L(\sigma_{A_n})$. Let $\mu = \mu_{A_1} \hat{\otimes} \cdots \hat{\otimes} \mu_{A_n}$ be the continuous homomorphism from the algebra $L(\sigma)$ to the algebra $End(B)$ which is made up from the μ_{A_i} in the theorem. For $f \in L(\sigma)$ denote $f(A_1, \ldots, A_n) = \mu(f)$. Then Vishik shows that $f(A_1, \ldots, A_n)$ is an analytic operator with compact spectrum

$\sigma_{f(A_1,\ldots,A_n)} \subset f(\sigma_{A_1},\ldots,\sigma_{A_n})$, and for $z \notin f(\sigma_{A_1},\ldots,\sigma_{A_n})$ we have

$$R_{f(A_1,\ldots,A_n)}(z) = \left(\mu(x_1,\ldots,x_n), \frac{1}{z - f(x_1,\ldots,x_n)}\right).$$

In addition, $\mu_{f(A_1,\ldots,A_n)} = f_*\mu$, i.e., if $\ell \in L(f(\sigma_{A_1},\ldots,\sigma_{A_n}))$, then

$$\mu_{f(A_1,\ldots,A_n)}(\ell) = \mu(\ell \circ f).$$

BIBLIOGRAPHY

1. W. Adams, Transcendental numbers in the p-adic domain, Amer. J. Math. 88 (1966), 279-308.

2. Y. Amice, Les nombres p-adiques, Presses Univ. de France, 1975.

3. Y. Amice, Interpolation p-adique et transformation de Mellin-Mazur, selon Hà huy Khoái, Groupe d'ét. d'anal. ultramét., 15 jan. 1979.

4. Y. Amice and J. Frèsnel, Fonctions zêta p-adiques des corps de nombres abéliens réels, Acta Arith. 20 (1972), 353-384.

5. Y. Amice and J. Vélu, Distributions p-adiques associées aux séries de Hecke, Journées arith., 1974.

6. Arithmetic Algebraic Geometry, Proceedings of a conference held at Purdue Univ., 1963.

7. E. Artin, The Gamma Function, New York: Holt, Rinehart & Winston, 1964.

8. J. Ax, On the units of an algebraic number field, Illinois J. Math. 9 (1965), 584-589.

9. A. Baker, Linear forms in the logarithms of algebraic numbers, Mathematika 13 (1966), 204-216.

10. D. Barsky, Transformation de Cauchy p-adique et algèbre d'Iwasawa, Math. Ann. 232 (1978), 255-266.

11. P. Berthelot, Cohomologie cristalline des schémas de caractéristique $p > 0$, Springer Lecture Notes in Math. 407 (1974).

12. P. Berthelot and A. Ogus, Notes on Crystalline Cohomology, Princeton Univ. Press, 1978.

13. Z. I. Borevich and I. R. Shafarevich, Number Theory, Academic Press, 1966.

14. M. Boyarsky, p-adic gamma functions and Dwork cohomology, Trans. A.M.S. 257 (1980), 359-369.

15. A. Brumer, On the units of algebraic number fields, Mathematika 14 (1967), 121-124.

16. S. Chowla and A. Selberg, On Epstein's zeta-function, J. Reine und angew. Math. 227 (1967), 86-110.

17. J. Coates, An effective p-adic analogue of a theorem of Thue, Acta Arith. 15 (1969), 279-305.

18. J. Coates and W. Sinnott, On p-adic L-functions over real quadratic fields, Inventiones Math. 25 (1974), 252-279.

19. P. Deligne, Valeurs de fonctions L et périodes d'intégrales, Proc. Symp. in Pure Math. 33 (1979), part 2, 313-346.

20. P. Deligne, Cycles de Hodge sur les variétés abéliennes, preprint.

21. P. Deligne and K. Ribet, Values of abelian L-functions at negative integers over totally real fields, I.H.E.S. preprint, October 1979.

22. J. Diamond, The p-adic log gamma function and p-adic Euler constants, Trans. A.M.S. 233 (1977), 321-337.

23. J. Diamond, On the values of p-adic L-functions at positive integers, Acta Arith. 35 (1979), 223-237.

24. N. Dunford and J. T. Schwartz, Linear Operators, 3 vols., Interscience, 1958, 1963, 1971.

25. B. Dwork, On the rationality of the zeta function of an algebraic variety, Amer. J. Math. 82 (1960), 631-648.

26. B. Dwork, On the zeta function of a hypersurface, Publ. Math. I.H.E.S. 12 (1962), 5-68.

27. B. Dwork, A deformation theory for the zeta function of a hypersurface, Proc. I.C.M. Stockholm 1962, 247-259.

28. B. Dwork, p-adic cycles, Publ. Math. I.H.E.S. 37 (1969), 327-415.

29. B. Ferrero and R. Greenberg, On the behavior of p-adic L-functions at s=0, Inventiones Math. 50 (1978), 91-102.

30. J. Frèsnel, Nombres de Bernoulli et fonctions L p-adiques, Ann. Inst. Fourier, Grenoble 17 (1967), 281-333.

31. R. Greenberg, On a certain ℓ-adic representation, Inventiones Math. 21 (1973), 117-124.

32. R. Greenberg, A generalization of Kummer's criterion, Inventiones Math. 21 (1973), 247-254.

33. B. H. Gross, On the periods of abelian integrals and a formula of Chowla and Selberg, Inventiones Math. 45 (1978), 193-211.

34. B. H. Gross, On an identity of Chowla and Selberg, J. Number Theory 11 (1979), 344-348.

35. B. H. Gross, On the behavior of p-adic L-series at s=0, to appear.

36. B. H. Gross, On the factorization of p-adic L-series, Inventiones Math. 57 (1980), 83-96.

37. B. H. Gross and N. Koblitz, Gauss sums and the p-adic Γ-function, Annals of Math. 109 (1979), 569-581.

38. A. Grothendieck, Formule de Lefschetz et rationalité des fonctions L, Sém. Bourbaki 1964/65, exp. 279.

39. Hà-huy-Khoái, p-adičeskaja interpoljacija i preobrazovanie Mellina-Mazura (p-adic interpolation and the Mellin-Mazur transform), Moscow dissertation, 1978.

40. R. Hartshorne, Algebraic Geometry, Springer-Verlag, 1977.

41. K. Iwasawa, Lectures on p-adic L-Functions, Princeton Univ. Press, 1972.

42. N. M. Katz, On the differential equations satisfied by period matrices, Publ. Math. I.H.E.S. 35 (1968), 71-106.

43. N. M. Katz, Travaux de Dwork, Sém. Bourbaki 1971/72, exp. 409, Springer Lecture Notes in Math. 317 (1973), 167-200.

44. N. M. Katz, p-adic properties of modular schemes and modular forms, Proc. 1972 Antwerp Summer School, Springer Lecture Notes in Math. 350 (1973), 70-189.

45. N. M. Katz, p-adic L-functions via moduli of elliptic curves, Proc. Symp. in Pure Math. 29 (1975), 479-506.

46. N. M. Katz, Higher congruences between modular forms, Annals of Math. 101 (1975), 332-367.

47. N. M. Katz, p-adic interpolation of real analytic Eisenstein series, Annals of Math. 104 (1976), 459-571.

48. N. M. Katz, The Eisenstein measure and p-adic interpolation, Amer. J. Math. 99 (1977), 238-311.

49. N. M. Katz, Formal groups and p-adic interpolation, Astérisque Soc. Math. de France 41-42 (1977), 55-65.

50. N. M. Katz, Lectures at Princeton University, Spring 1978.

51. N. M. Katz, p-adic L-functions for CM fields, Inventiones Math. 49 (1978), 199-297.

52. N. M. Katz, Crystalline cohomology, Dieudonné modules, and Jacobi sums, to appear.

53. N. Koblitz, p-adic Numbers, p-adic Analysis, and Zeta-Functions, Springer-Verlag, 1977.

54. N. Koblitz, Gamma function identities and elliptic differentials on Fermat curves, Duke Math. J. 45 (1978), 87-99.

55. N. Koblitz, Interpretation of the p-adic log gamma function and Euler constants using the Bernoulli measure, Trans. A.M.S. 242 (1978), 261-269.

56. N. Koblitz, A new proof of certain formulas for p-adic L-functions, Duke Math. J. 46 (1979), 455-468.

57. M. Krasner, Rapport sur le prologement analytique dans les corps values complets par la méthode des éléments analytiques quasi-connexes, Bull. Soc. Math. France, Mém. 39-40 (1974), 131-254.

58. T. Kubota and H. W. Leopoldt, Eine p-adische theorie der zetawerte I, J. Reine und angew. Math. 214/215 (1964), 328-339.

59. S. Lang, Algebraic Number Theory, Addison-Wesley, 1970.

60. S. Lang, Introduction to Modular Forms, Springer-Verlag, 1976.

61. S. Lang, Cyclotomic Fields, Springer-Verlag, 1978.

62. S. Lang, Cyclotomic Fields, vol. 2, Springer-Verlag, 1980.

63. H. W. Leopoldt, Zur Arithmetic in abelschen zahlkörpern, J. Reine und angew. Math. 209 (1962), 54-71.

64. H. W. Leopoldt, Eine p-adische theorie der zetawerte II, J. Reine und angew. Math. 274/275 (1975), 224-239.

65. E. Lindelöf, Le calcul des résidus, Paris: Gauthier-Villars, 1905.

66. K. Mahler, An interpolation series for a continuous function of a p-adic variable, J. Reine und angew. Math. 199 (1958), 23-34.

67. Ju. I. Manin, O matrice Hasse-Vitta algebraičeskoǐ krivoǐ (On the Hasse-Witt matrix of an algebraic curve), Izvestija AN SSSR 25 (1961), 153-172.

68. Ju. I. Manin, Periodi paraboličeskih form i p-adičeskie rjady

Gekke (Periods of cusp forms and p-adic Hecke series), Mat. Sb. 93 (1973), 378-401.

69. Ju. I. Manin, Ne-arhimedovo integrirovanie i p-adičeskie L-funkcii Žake-Lenglendsa (Non-archimedean integration and p-adic Jacquet-Langlands L-functions), Uspehi Mat. Nauk 31 (1976), 5-54.

70. Ju. I. Manin and M. M. Vishik, p-adičeskie rjady Gekke dlja mnimyh kvadratičnyh poleĭ (p-adic Hecke series for quadratic imaginary fields), Mat. Sb. 95 (1974), 357-383.

71. B. Mazur, Analyse p-adique, Bourbaki report (unpublished), 1972.

72. B. Mazur and H. P. F. Swinnerton-Dyer, Arithmetic of Weil curves, Inventiones Math. 25 (1974), 1-61.

73. H. Minkowski, Zur Theorie der Einheiten in den Algebraische Zahlkörpern, Göttingen Nachrichten, 1900.

74. P. Monsky, p-adic Analysis and Zeta Functions, Lectures at Kyoto Univ., Kinokuniya Book Store, Tokyo, or Brandeis Univ. Math. Dept., 1970.

75. P. Monsky and G. Washnitzer, Formal cohomology I, II, III, Annals of Math. 88 (1968), 181-217; 88 (1968), 218-238; 93 (1971), 315-343.

76. Y. Morita, A p-adic analogue of the Γ-function, J. Fac. Sci. Univ. Tokyo 22 (1975), 255-266.

77. N. Nielsen, Die Gammafunktion, Chelsea Publ. Co., 1965.

78. Ju. V. Osipov, O p-adičeskih dzeta-funkcijah (On p-adic zeta-functions), Uspehi Mat. Nauk 34 (1979), No. 3, 209-210.

79. G. Overholtzer, Sum functions in elementary p-adic analysis, Amer. J. Math. 74 (1952), 332-346.

80. F. Pham, Formules de Picard-Lefschetz généralisées et ramification des intégrales, Bull. Soc. Math. France 93 (1965), 333-367.

81. K. Ribet, p-adic interpolation via Hilbert modular forms, Proc. Symp. in Pure Math. 29 (1975), 581-592.

82. J.-P. Serre, Endomorphismes complètement continus d'espaces de Banach p-adiques, Publ. Math. I.H.E.S. 12 (1962), 69-85.

83. J.-P. Serre, Dépendance d'exponentielles p-adiques, Sém. Delange-Pisot-Poitou, 7 année, exp. 15.

84. J.-P. Serre, A Course in Arithmetic, Springer-Verlag, 1973.

85. J.-P. Serre, Formes modulaires et fonctions zêta p-adiques, Proc. 1972 Antwerp Summer School, Springer Lecture Notes in Math. 350 (1973), 191-268.

86. J.-P. Serre, Sur le résidu de la fonction zêta p-adique d'un corps de nombres, C.R. Acad. Sci. Paris 287 (1978), 183-188.

87. I. R. Shafarevich, Basic Algebraic Geometry, Springer-Verlag, 1974.

88. L. G. Shnirel'man, O funkcijah v normirovannyh algebraičeski zamknutyh telah (On functions in normed algebraically closed division rings), Izvestija AN SSSR 2 (1938), 487-498.

89. C. Siegel, Über die Fourierschen koeffizienten von modulformen, Göttingen Nachrichten 3 (1970), 15-56.

90. H. Stark, L-functions at s=1, II and III, Advances in Math. 17 (1975), 60-92 and 22 (1976), 64-84.

91. J. Tate, Rigid analytic spaces, Inventiones Math. 12 (1971), 257-289.

92. J. Tate, Arithmetic of elliptic curves, Inventiones Math. 23 (1974), 179-206.

93. J. Tate, On Stark's conjecture, to appear.

94. M. M. Vishik, Ne-arhimedovy mery svjazannye s rjadami Dirihle (Non-archimedean measures connected with Dirichlet series), Mat. Sb. 99 (1976), 248-260.

95. M. M. Vishik, Nekotorye primenenija integrala Shnirel'mana v ne-arhimedovom analize (Some applications of the Shnirelman integral in non-archimedean analysis), preprint.

96. M. M. Vishik, O primenenijah integrala Shnirel'mana v nearhimedovom analize (On applications of the Shnirelman integral in non-archimedean analysis), Uspehi Mat. Nauk 34 (1979), 223-224.

97. G. N. Watson and E. T. Whittaker, A Course of Modern Analysis, 4th ed., Cambridge Univ. Press, 1927.

98. A. Weil, Sur les courbes algébriques et les variétés qui s'en déduisent, Paris: Hermann, 1948.

99. A. Weil, Numbers of solutions of equations in finite fields, Bull. A.M.S. 55 (1949), 497-508.

100. A. Weil, Jacobi sums as Grössencharaktere, Trans. A.M.S. 73 (1952), 487-495.

101. A. Weil, Sur les périodes des intégrales abéliennes, Comm. on Pure and Appl. Math. 29 (1976), 813-819.

102. D. Widder, The Laplace Transform, Princeton Univ. Press, 1941.

INDEX